T0269141

CAMBRIDGE LIBRARY COLLECTION

Books of enduring scholarly value

Physical Sciences

From ancient times, humans have tried to understand the workings of the world around them. The roots of modern physical science go back to the very earliest mechanical devices such as levers and rollers, the mixing of paints and dyes, and the importance of the heavenly bodies in early religious observance and navigation. The physical sciences as we know them today began to emerge as independent academic subjects during the early modern period, in the work of Newton and other 'natural philosophers', and numerous sub-disciplines developed during the centuries that followed. This part of the Cambridge Library Collection is devoted to landmark publications in this area which will be of interest to historians of science concerned with individual scientists, particular discoveries, and advances in scientific method, or with the establishment and development of scientific institutions around the world.

Elements of Electricity and Electro-Chemistry

The amateur scientist George John Singer (1786–1817) worked in the family business of artificial flower and feather making, but all his spare time was absorbed in the study of electricity and electromagnetism. He invented his own apparatus, including a gold-leaf electrometer, and built a laboratory-cum-lecture room at the back of his house: his public demonstrations were attended by Faraday and Francis Ronalds, and he was also a friend of the pioneering 'electrician' Andrew Crosse. This significant book, published in 1814, demonstrates the breadth of Singer's knowledge of his subject and of other contemporary work in the field. It describes in detail electric phenomena, in nature and in the laboratory, covering a wide range of experiments with and applications of electricity, and discussing the work of Franklin, Volta, Crosse and Dalton, among others. Sadly, Singer's promising scientific career was brought to an early end by tuberculosis: he died aged only thirty-one.

Cambridge University Press has long been a pioneer in the reissuing of out-of-print titles from its own backlist, producing digital reprints of books that are still sought after by scholars and students but could not be reprinted economically using traditional technology. The Cambridge Library Collection extends this activity to a wider range of books which are still of importance to researchers and professionals, either for the source material they contain, or as landmarks in the history of their academic discipline.

Drawing from the world-renowned collections in the Cambridge University Library and other partner libraries, and guided by the advice of experts in each subject area, Cambridge University Press is using state-of-the-art scanning machines in its own Printing House to capture the content of each book selected for inclusion. The files are processed to give a consistently clear, crisp image, and the books finished to the high quality standard for which the Press is recognised around the world. The latest print-on-demand technology ensures that the books will remain available indefinitely, and that orders for single or multiple copies can quickly be supplied.

The Cambridge Library Collection brings back to life books of enduring scholarly value (including out-of-copyright works originally issued by other publishers) across a wide range of disciplines in the humanities and social sciences and in science and technology.

Elements of Electricity and Electro-Chemistry

George John Singer

CAMBRIDGE
UNIVERSITY PRESS

University Printing House, Cambridge, CB2 8BS, United Kingdom

Cambridge University Press is part of the University of Cambridge.
It furthers the University's mission by disseminating knowledge in the pursuit of education, learning and research at the highest international levels of excellence.

www.cambridge.org
Information on this title: www.cambridge.org/9781108079228

This edition first published 1814
This digitally printed version 2015

ISBN 978-1-108-07922-8 Paperback

ELEMENTS

OF

ELECTRICITY

AND

ELECTRO-CHEMISTRY.

By GEORGE JOHN SINGER.

LONDON:

PRINTED FOR LONGMAN, HURST, REES, ORME, AND
BROWN, PATERNOSTER ROW;
AND R. TRIPHOOK, ST. JAMES'S STREET.

1814.

T. BENSLEY, PRINTER,
Bolt Court, Fleet Street.

ERRATA.

Table of Contents, p. xx, l. 12, *for* various part, *read* various parts.
Introduction, p. 11, l. 22, *for* Muschenbrouk, *read* Muschenbroek.
Page 98, line 9, *for* their lamina, *read* thin lamina.
192, line 1, *for* fig. 24, *read* fig. 25.
246, line 23, *for* thinks its, *read* thinks it.
247, line 9, *for* its form, *read* its nature.
256, line 14, *for* iles, *read* Isles.
417, line 13, *for* second, *read* third.
447, line 24, *for* Electrio, *read* Electro.

*** The various new and improved articles of electrical apparatus described in this work may be obtained, of the most perfect construction, at No. 3, Princes Street, Cavendish Square, London.

PREPARING FOR THE PRESS,

BY THE SAME AUTHOR,

A POPULAR VIEW

OF THE

PHENOMENA OF THE ATMOSPHERE;

INCLUDING

An Account of the Instruments and Observations most essential in Meteorology,

AND AS INDICATIONS OF APPROACHING CHANGES OF WEATHER.

WITH OBSERVATIONS ON THE ECONOMY OF LIGHT AND HEAT.

In one volume octavo, with Plates.

PREFACE.

THE Science of which a correct outline is attempted in the following pages, has been the subject of occasional consideration for upwards of two centuries, but has rarely been the object of earnest and complete attention. The surprizing nature of its phenomena, and the ease with which many of them are produced, have made it in the hands of the ignorant, and the empirical, a mere vehicle of shew and deception.

The attention of philosophers has indeed been occasionally directed to Electricity, as a promising source of discovery; and after the immortal Franklin had demonstrated its con-

nexion with atmospherical phenomena, it was always regarded as an important natural agent. But the insulated nature of many of the observed facts, and the difficulty of referring them to any simple general principle, continued to withhold from it that active and persevering cultivation which is essential to the promotion, and correct application of every branch of science.

The discovery of the electrical powers of the Torpedo, and the Gymnotus, the researches of Galvani on Animal Electricity, the invention of the Voltaic Battery, and the discoveries with which it has so recently enriched and diversified Chemical Philosophy, have amply demonstrated the connection of Electricity with the most interesting subjects of experimental enquiry. Its importance is therefore sufficiently obvious, and no obstacle is opposed to its extensive cultivation but the present difficulty of attaining a competent acquaintance with the various facts that have been already ascertained, and a just conception of their mutual relation and dependance.

There are indeed many useful works on Electricity, but in the present extended state of the Science a proper acquaintance with its principles can only be obtained by the laborious collection of its widely scattered materials; which it will then require considerable attention to arrange, and tolerable skill to combine; hence the appropriation of a portion of time is required for this purpose, which is incompatible with the pursuits of many active and intelligent enquirers, and will be generally considered as inconvenient and objectionable.

To remove this difficulty is the principal object of the present work; and the devotion of a considerable portion of my life to the practical consideration of the subject, with the constant habit of explanation and arrangement, in Lectures, authorizes an expectation that the attempt may be successful.

By a proper attention to arrangement, I have been enabled to communicate a more extensive collection of facts in a single volume, than is to be found in any existing treatise with which I am acquainted; and that attention may be expected to render even more concise statements amply intelligible; for materials thus disposed, are like the combination of stones in an Arch; they mutually support each other, and form a connected series, in which every part is essential to the existence of the whole.

The experiments described, are sufficiently numerous, and some of them original; the simplicity by which many of the most essential are characterized, will be advantageous to the student who may not have access to a regular apparatus, or to proper experimental lectures; and it will be of advantage to every one, to repeat as many of the described experiments as possible, during the regular perusal of the work.

Some apology may be necessary for the free use of chemical terms; but this could not have been avoided without omitting a most essential part of the subject, and a sufficient explanation of them, would have been too extensive for the limits of an elementary treatise. Those who are unacquainted with Chemistry, may prepare themselves for this part of the subject, by an attentive perusal of any of the very excellent works on that science, of which we have at present such numerous examples.

CONTENTS.

INTRODUCTION.

PART I.

ON ELECTRICAL PHENOMENA, AND THE CIRCUMSTANCES ESSENTIAL TO THEIR PRODUCTION.

CHAP. I.

Nature of Electrical Action, and Sources of Electrical Excitation.—Positive and Negative Electricity.

CHAP. II.

Of Conductors and Nonconductors of Electricity, and of the Electrical Apparatus.

CHAP. III.

Experiments with the Electrical Machine.—Theory of its Action.—Phenomena of Attraction and Recession.

CHAP. IV.

On the Phenomena of Electric Light.

CHAP. V.

On the Leyden Jar, and the Nature of Electrical Influence.

PART II.

OF THE MECHANICAL AND CHEMICAL AGENCIES OF ELECTRICITY.

CHAP. I.

Instruments required for the Application of the Electric Power to the Purpose of Experiment.

CHAP. II.

Mechanical Effects of Electricity.

CHAP. III.

Chemical Effects of Electricity.

PART III.

NATURAL AGENCIES OF ELECTRICITY.

CHAP. I.

On the Identity of Electricity, and the Cause of Lightning.

CHAP. II.

On the Phenomena of Thunder Storms, and the probable Sources of Atmospherical Electricity.

CHAP. III.

On some luminous Phenomena of the Atmosphere, the Observation of Atmospherical Electricity, and the Arrangement of a new System of Insulation.

CHAP. IV.

Connection of Electricity with Medicine, and Natural History.

PART IV.

VOLTAIC ELECTRICITY.

CHAP. I.

Structure of the Voltaic Apparatus, and Nature of its Electrical Phenomena.

CHAP. II.

Chemical Effects of the Voltaic Apparatus.

CHAP. III.

Extensive Agency of the Voltaic Apparatus, as an Instrument of Analysis.—Its Influence in the Evolution of Light and the Production of Heat.

CHAP. IV.

Sketch of the State of Theoretical Knowledge in Voltaic Electricity.—Structure and Properties of the Electric Column.

APPENDIX.

VARIOUS ADDITIONS AND CORRECTIONS.—EXPERIMENTS ON THE ELECTRICAL EFFECTS EVINCED AFTER THE CONTACT OF DIFFERENT BODIES.

INTRODUCTION.

At the commencement of the 17th century, a new branch of natural science was created by the experiments and inquiries of an English physician, William Gilbert; who, in a Latin Treatise, "De Magnete;" published in the year 1600, described the existence of an attractive power as the effect of friction on various bodies. This property had been observed by the ancients, as peculiar to the well-known mineral production Amber; and hence all the substances enumerated by Gilbert and others, as possessed of analogous powers, have been called Electrics; and the unknown cause of such phenomena, Electricity.*

During the whole of the 17th century, but little addition was made to the discoveries of Gilbert; his Catalogue of Electrics was extended by the celebrated Boyle, who discovered

* From ηλεκτρον, the Greek term for amber.

that their attractive power was much increased, by warming, and wiping them, before they were subjected to friction; in such cases he observed, that the electrics, whilst rubbing, frequently emitted flashes of light, and he considered this appearance as an additional characteristic of the electric power.

The discoveries of Boyle were confirmed by his contemporary Otto Guericke (inventor of the Air Pump.) This philosopher constructed an apparatus in which the electric was made to revolve, as a more convenient mode of applying friction. His instrument was the same in principle as that now familiarly known as the Electric Machine. By its aid he ascertained the constant appearance of light as an attendant on strong electric excitation, and discovered the curious fact, that electric attraction is generally followed by apparent repulsion.

In the year 1675, Sir Isaac Newton discovered that electric attraction was not prevented by the interposition of a plate of glass; the opposite side to that which has been rubbed being also capable of attracting light bodies; and exciting some curious varieties of motion in them.

At the commencement of the 18th century, the first Treatise on Electricity was published by Mr. Hauksbee: it contains an account of all the facts ascertained by his predecessors; and a variety of new experiments, made principally to ascertain the nature of electric light. His discoveries were numerous and important, but scarcely of sufficient magnitude to constitute a distinct epoch in the science. His most singular discovery was the great facility with which the electric light is produced in a vacuum.

Toward the year 1729, an important discovery was made by Mr. Stephen Grey, a pensioner of the Charter-house, who at that time cultivated this then infant science with great industry and address. Directing his attention to the nature of electric phenomena, he endeavoured to excite them in all known bodies, and by this means extended very considerably the catalogue of electrics; many substances in which no attractive power was excited by friction whilst in their natural state, became strongly attractive if rubbed after they had been moderately heated, but lost this faculty sooner or later when cooled. This fact clearly pointed out a relation between the *state* of bodies and their power of evincing electric appearances; and the nature of this relation was explained by Mr. Grey's subsequent experiments. Every attempt to render metals electric by friction had proved ineffectual in the hands of Mr. Grey, as well as in those of preceding inquirers, when it occurred to him, that as electric light appeared to pass between excited bodies and such as were incapable of excitation, the attractive power might be also capable of communication from one to the other. He inserted a wire and ball, by means of cork, in the extremity of a glass tube, and on rubbing the tube, found its attractive power was communicated to the wire and ball. He employed longer wires, till their vibration prevented him from extending them further. He then suspended the ball by means of pack-thread, from the tube; the electricity was still communicated. He ascended a balcony twenty-six feet high, and suspending the ball from his tube by a proportional length of string, found that the electricity was communicated from the tube to it, so as to attract light bodies from the pavement of the court below. Associated with Mr. Wheeler, Mr. Grey afterwards extended his experiments, and in one instance transmitted the attractive power of his excited tube through nearly 800 feet of packthread, without any apparent diminution of its force. In arranging the apparatus for these experiments, it was found that a *silk* line was incapable of transmitting the attractive power of the tube; an effect which these experimenters at first attributed to its comparative smallness; but they afterwards observed that a *wire* of much smaller diameter conveyed the electric effect completely, and thus discovered that there are in nature various bodies differently fitted for the transmission of electricity, some conveying it with facility to a great extent, and others apparently unsuited to transmit it to any perceptible distance. The first class of bodies are now called Conductors of Electricity, and the second class Non-Conductors, or Electrics: terms which appear to have been first proposed by Dr. Desaguliers.

The experiments of Messrs. Grey and Wheeler show that conducting power does not depend on the magnitude, but on some peculiarity in the nature of bodies; a peculiarity whose cause has not yet been discovered.

In 1732, M. Du Faye, Member of the Academy of Sciences at Paris, repeated and extended the experiments of Mr. Grey: he ascertained that the conducting power of packthread and other vegetable and animal substances is principally dependant on the water they contain: he conveyed electricity to greater distances by wetting the packthread, and found the conducting faculty became less perfect in most fibrous bodies in proportion as their natural moisture was expelled. He also observed, that such substances as were least susceptible of electric excitement by friction were the best conductors of electricity; though all the bodies he tried became electric by communication when placed on a non-conducting support. In this way he electrified himself, being supported by silk lines, and touched by an excited glass tube; and on this occasion the Abbé Nollet, who accompanied him in these experiments, drew the first electrical spark from the human body.

M. Du Faye has also the merit of having given the first clear account of that apparent repulsion which obtains in most electric experiments, and was first observed by Otto Guericke, who had noticed that the fibres of an electrified feather receded from each other, and from the tube or globe with which they had been electrified. Du Faye viewed this phenomenon as the indication of a general principle in electricity, which may be thus expressed. Electrified bodies attract all those which are not so, but repel them as soon as they are electrified by their contact. Thus leaf gold is first attracted by an excited tube brought near to it, becomes electrical by the contact, and is then repelled; nor will it be again attracted while it retains its electric quality: but if it come in contact with any unelectrified body it loses its electricity, and will be again attracted by the excited tube; until, electrified by it, it is again repelled: and thus may alternate attraction and repulsion be produced as long as the excited tube retains its power.*

The consideration of this general principle led the same assiduous philosopher to a discovery of the first importance, viz. the existence of two distinct attractive powers, produced by the friction of different substances. The one,

* This principle is here stated nearly in the language of Du Faye, and of most subsequent writers. Its propriety will be considered hereafter.

excited by rubbing glass, rock crystal, gems, wool, hair, and many other substances, he called Vitreous Electricity. The other, resulting from the friction of amber, copal, gum-lac, resins, sealing-wax, &c. he named Resinous Electricity. The characteristics of these attractive powers are, that they strongly attract each other, and produce a mutual counteraction of effect, whilst they separately act in an apparently similar manner on all unelectrified bodies: but the effect of either of them is destroyed or weakened by the approach of the other. If gold leaf be electrified by rubbed glass it immediately recedes from it, and will not again approach whilst it retains its electric state. But in this state it is strongly attracted by any excited body of the resinous class, and will fly to sealing-wax or amber more rapidly than to an unelectrified body. Hence it was concluded by Du Faye, that there are two distinct electricities, each repulsive of its own particles, but having a strong attraction for those of the other. So that all bodies electrified with the vitreous electricity repel those that are similarly electrified, and attract such as are unelectrified or endowed with the resinous electricity.

And the converse of this is the case with such as are possessed of the resinous electricity.

The terms resinous and vitreous electricity, were sufficiently appropriate at the time they were proposed; but it has been since found, that either electricity can be obtained at pleasure, both from glass and sealing-wax, by varying the nature of the substance with which they are rubbed. Hence the vitreous electricity of Du Faye is now called positive electricity; and the resinous, negative electricity; terms first proposed by Dr. Franklin.

To the labours of Messrs. Grey and Wheeler, and their coadjutors Du Faye and Nollet, all subsequent electricians are highly indebted; their means of research were extended by the improvement of electrical apparatus, necessarily resulting from the discovery of conducting and non-conducting power; whilst the generalization of electric phenomena by Du Faye, and his discovery of the distinction between positive and negative electricity, was an enlargement of the existing sphere of knowledge in a degree before unparalleled. From this period, indeed, the science assumed a more important aspect, its cultivators increased in number, and the communication of their researches constituted a prominent feature in the transactions of the most celebrated societies and academies of Europe.

In 1742, the Germans began to distinguish themselves by an active attention to electricity; they improved very considerably the electrical apparatus, applying the principles discovered by Grey to the perfection of the known sources of electrical excitation. To them we owe the employment of a cushion instead of the hand, as a means of applying friction to excite electricity: the idea of applying multiplying wheels, as a method of accelerating in any proportion the rapidity of such friction; the cylindrical form of the electric machine; and the employment of an insulated conductor to concentrate the power of the excited electric, and apply it more conveniently to experiment.

The great power of the apparatus contrived by the Germans, increased the analogy before observed between the appearance of electric light and common fire; and induced them to apply it more extensively to the purpose of experiment. They succeeded by its means in inflaming ether, spirits of wine, and other inflammable bodies; and thus were the first to demonstrate its possible application as a chemical agent. Dr. Watson and some other members of the Royal Society pursued the path of discovery opened by these researches; and the singularity and novelty of the effects produced, attracted very general attention to a subject, which was soon destined to excite universal admiration and inquiry.

In the year 1746, a discovery was made by some professors in the university of Leyden, of a method of accumulating the electric power to an extraordinary degree. The experiment consisted in enclosing some water or other conducting substance in a glass vessel, and electrifying it; if then the outside of the glass vessel was grasped with one hand, and the enclosed conductor or any substance connected with it touched with the other, a bright spark ensued, and a violent convulsive motion was felt in the arms and across the breast. Professor Muschenbrouk, Messrs. Cuneus, Alemand, and Winkler, made the experiment with water in glass jars or bottles; and M. Von Kliest (who it is said first made the discovery) employed a phial in which a blunt piece of wire was loosely placed. This experiment soon became popular, the apparatus received the name of Leyden Jar, or Leyden Phial, and the sensation it produced the Electric Shock. The first experimenters gave ludicrous and exaggerated accounts of its effects, and to this circumstance may perhaps be partly attributed the public curiosity it so promptly and highly excited. In the same year it was shewn by itinerant exhibitors in almost every part of Europe; and the experiment was repeated and varied by the electricians of every country.

To trace the progress of electrical discovery beyond this period in chronological succession, would be incompatible with the limit of an elementary work; and, from the extent and variety of succeeding inquiries, even less eligible for a detailed history than a connected arrangement of the discoveries in the order of their mutual dependance; a plan which it is my intention to adopt, at no very distant period, in a comprehensive history of this science, from its origin to the present time.

To complete the preceding sketch, as a general outline of the science, I shall now briefly enumerate the most important electrical discoveries that have succeeded the Leyden experiment. Dr. Watson in England, and Dr. Franklin at Philadelphia, nearly at the same time, and without any previous concert, improved the structure of the Leyden jar, and advanced an explanation of the various phenomena of electricity. Their propositions were nearly similar; but that of Dr. Franklin being most perfect, and having a real priority of publication, was adopted, and has been since celebrated as the Franklinean Theory of Electricity. He referred all electrical effects to the motion of a peculiar fluid, repulsive of its own particles, and having an attraction for all other matter. And he considered the opposite electricities of glass and sealing-wax as indications of different states of this fluid: the vitreous electricity being the plus or positive state, and the resinous the minus or negative state. All bodies can contain a certain quantity of electric fluid in a latent state. If this quantity be increased they become electrified positively; if it be diminished they are rendered negative. The production of electrical effects is therefore nothing but the result of the unequal distribution, by art, of a naturally diffused fluid.

Such are the leading principles of the Franklinean theory; they have been considered mathematically by Mr. Cavendish, and by Æpinus, and, with some modifications, apply to most of the phenomena at present known.

Our existing theoretical views are derived principally from this source, aided by the researches of Messrs. Canton, Kinnersly, Henly, Beccaria, Cavallo, Bennet, Volta, DeLuc, Morgan, Cuthbertson, &c. who have severally contributed most extensively to this branch of knowledge.

Before the discovery of the Leyden Jar, electricity had not been applied to any useful purpose. Dr. Franklin was the first to supply this deficiency; he pointed out a striking resemblance in the effects of lightning and electricity; and, conceiving they might result from different modifications of the same power, proposed to verify his conjecture by experiment. He conceived the bold attempt of collecting lightning from the clouds, and subjecting it to examination. His suggestion was first attended to by the French philosophers: they confirmed his conjecture on the 10th of May, 1752, and performed the ordinary experiments in electricity by means of that terrific power. Before any account of their success had reached Dr. Franklin, he had himself obtained a similar result; and the experiment was soon repeated in almost every civilized country.* Franklin immediately followed the confirmation of his discovery, by its application to the defence of buildings and ships from injury by thunder storms; and his endeavours were attended by every expected success. They were followed by the observations of other able philosophers, and will (with them) be considered in a subsequent part of this work.

The discovery of the atmospherical agency of electricity drew the attention of philosophers more closely to that subject, and their inquiries have for the last sixty years been attended by continued accessions of knowledge; many other sciences have been enlightened and assisted by its cultivation; and the more refined natural phenomena are rarely investigated without the cooperation of its principles.

Electricity has been applied to medicine with various success; much of empiricism

* It will be described under the section on the natural agencies of electricity.

clouded its earlier application; for mysterious agents are always convenient implements of imposture; but the delusions of quackery, and the mistakes of ignorance, ought not to operate against the scientific employment of a powerful agent, so easily controuled and susceptible of such various application. At present, the medical employment of electricity is most frequently adopted when other remedies have failed; and its success, under such circumstances, is a cogent argument for more extensive and impartial trials.

The connection of electricity with natural history has been demonstrated by Mr. Walsh, and Mr. Cavendish. They have shewn that the torpedo and gymnotus owe their extraordinary power of benumbing the faculties of animals of superior strength, and of arresting the progress of the swiftest of the watery tribe, to the faculty of suddenly accumulating electricity, and of discharging it in any direction at pleasure. The effect of electricity on the animal economy has been an object of attention at intervals, since the first observation of the peculiar sensation of the electric shock. Beccaria of Turin, appears to have first noticed the power of electricity in producing muscular motion. And Galvani, in 1791, instituted a series of experiments which promised to lead to a full developement of the nature of the nervous influence, and of the origin of muscular motion. His hypothesis, which supposed the constant agency of electricity in the animal economy, was opposed by professor Volta, who had already distinguished himself by important electrical discoveries. This controversy gave rise to the invention of a new source of electric power, the Voltaic Battery; an instrument which, in the short space of thirteen years, has created a new science, and effected the most important discoveries in chemical philosophy.

The application of electricity, as a chemical agent, was, as I have already stated, first suggested by the experiments of the Germans.

It was successively employed in this way by Dr. Watson, the Abbé Nollet, Dr. Franklin, Mr. Kinnersly, Signior Beccaria, and most extensively by Dr. Priestley. The inquiry was pursued by Mr. Cavendish, Professor Volta, the French academicians, a society of Dutch chemists, Mr. Cuthbertson, Mr. Morgan, and Dr. Pearson. I shall endeavour to notice the most important of their experiments, as the subjects to which they relate shall occur.

During the interval that elapsed from the discovery of the Leyden Jar, to the invention of the Voltaic Battery, many important practical improvements had been made in the methods of experiment. The original apparatus was greatly improved, and many new instruments invented. Dr. Priestley, Mr. Nairne, and Mr. Cuthbertson, successively contributed to the perfection of the electrical machine, and the apparatus for accumulating its power, or directing it to the purpose of experiment. And for the more refined purposes of inquiry, various delicate tests of electrical action were contrived by Canton, Cavallo, Bennett, Volta, Nicholson, Read, and Cuthbertson. To these instruments we are indebted for some of the most interesting discoveries.

From the period of the invention of the Voltaic Battery, the progress of experiment has been so intimately blended with chemical inquiry, as to constitute a distinct branch of science. This part of the subject originating with the discovery of Volta, and continued by the researches of the most celebrated philosophers of modern times, I have thought it necessary to arrange, and consider separately under the general term, Voltaic Electricity.

ELEMENTS

OF

ELECTRICITY.

PART I.

OF ELECTRICAL PHENOMENA, AND THE CIRCUMSTANCES ESSENTIAL TO THEIR PRODUCTION.

CHAP. I.

Nature of Electrical Action, and Sources of Electrical Excitation.—Positive and Negative Electricity.

If a white and a black silk ribbon, of two or three feet long, perfectly dry, be applied to each other by their flat surfaces, and are then drawn repeatedly between the finger and thumb, or over dry silk-velvet, or woollen cloth, they will be found to adhere to each other; and if separated at one end will rush together again with great rapidity. Each ribbon, when separated, will attract any light substances to which it is presented; and if the experiment be made in a dark room, a flash of light will occasionally attend the separation of the ribbons.

Sticks of sealing-wax, resin, or sulphur, when rubbed with dry woollen cloth, or fur; and tubes or rods of glass, when rubbed with silk, exhibit similar powers; and if of sufficient size, produce, when applied within a short distance of the face or hand, a distinct and singular sensation.

These effects having been first produced by the friction of amber (electron), are called electrical phenomena; and the processes employed for their production, the excitation of electricity.

Attraction is the phenomenon most constantly attendant on excitation; it is therefore considered as an indication of the presence and action of electricity, and is the basis of all its tests. Electricians formerly, for such trials, employed a light wooden or metal needle, supported by its centre on a point, or a thread or feather delicately suspended. To these the excited body was presented, and if they were attracted by it, the attraction was attributed to electricity, and the excited body called an electric.

The suspended needle, and every contrivance employed for the same purpose, is called an electroscope, when designed to indicate the existence of electrical phenomena; and an electrometer, when considered as a measure of their force. The latter term alone appears necessary; for every contrivance hitherto employed to ascertain the presence of electrical phenomena, is also calculated to measure their power. Thus in the instance of the ballanced needle, or suspended feather; the greater or less distance to which it is necessary to bring the excited body before attraction ensues, supplies a measure of the force of such attraction.

The most useful electrometers are constructed by suspending two slips of gold leaf from the cap of a glass cylinder, as represented by Fig. 1. The slips of gold leaf hang in the centre of the cylinder, parallel and contiguous, when unelectrified; but separate from each other (as shewn in the figure) when electrified;

in consequence of their attraction for the surrounding air and the sides of the glass.*

Small balls turned from the pith of elder, and suspended by fine threads or silver wires, are sometimes substituted for the gold strips. They are less easily affected, but they are more durable. The pith balls suspended by thread or wire, are also occasionally used without a glass cylinder.—See Fig. 2.

Electrical phenomena then are characterized by the attraction and recession of light substances; the consequent production of motion in them, and of sensation in living bodies, and by the evolution or production of light. Various are the means by which these effects may be

* The separation of electrified bodies is usually ascribed to repulsion; an assumption quite hypothetic and unnecessary. Lord Stanhope has shewn that the separation is *less* in *rare* than in *dense* air; which is contrary to what ought to obtain if repulsion was its cause. Mr. Kinnersley first remarked that there were no proofs of electric repulsion, and shewed that the supposition of its existence was a barrier to the explanation of electric phenomena. See Franklin's Electricity, page 384.—Mr. G. Morgan has since expressed the same opinion; which is also entertained by professor Volta. It occurred to me before I knew of these authorities, and would, I should presume, to any one sufficiently acquaint d with the varieties of electrical action.

produced, but their most obvious sources are the following:

1st. Friction.

2d. Change of form.

3d. Change of temperature.

4th. Contact of dissimilar bodies.

The instances of the first kind are most numerous, and indeed under certain limits universal. They may be obtained by rubbing any one of a most extensive list of resinous and silicious substances; and of dry, vegetable, animal, and mineral productions. The electricity thus excited, is most readily shewn, by presenting the rubbed substance to the cap of the gold-leaf electrometer.

Examples of the second kind are also extensive. Sulphur melted and poured into a conical wine glass, contracts, and becomes electrical in cooling. A silk thread, or a stick of glass should be inserted in the sulphur whilst fluid, to serve as a handle; when cold it may easily be separated from the glass by its handle, and will then affect the electrometer, and evince other electric signs. If the sulphur cone be kept in the glass in which it was made, it will preserve its electric power for years, and evince them perceptibly, whenever the glass and sulphur are separated.

Chocolate, when it congeals after fusion, exhibits similar properties; and Chaptal observed the same circumstance during the congelation of glacial phosphoric acid. Calomel also, when it fixes by sublimation to the upper part of a glass vessel, has been found strongly electrical. The condensation of vapour, and the evaporation of fluids, though apparently opposite processes, are alike sources of electrical excitation.

Various crystallized gems, and a stone called the Tourmalin, become electrical by the mere application of heat; but no other substances have yet unequivocally manifested the same property; though the effects of friction are generally increased, if it is preceded by a moderate elevation of temperature.

The contact of dissimilar bodies is probably in all cases the real primary cause of electrical excitement, but it is rarely employed alone; for electricity is known to us only by its effects, which are constantly the result of an artificial arrangement, and consequently may not immediately succeed the primary cause of electric motion. Muscovy talc, when its lamina are suddenly torn apart, appears electrical, and sometimes exhibits a bright flash of light; such is occasionally the case with other substances, but we have no evidence from these experiments that will enable us to decide whether separation be the primary, or only the proximate cause of the phenomena. In most instances contact appears to *produce* an effect, which is to be *exhibited* only by separation; but there is one decisive instance of electricity produced by contact alone, the electric column, (invented by J. A. De Luc, esq.) It consists of 800 or 1000 small discs, of silver, zinc, and paper; placed upon each other regularly, in the order named, and enclosed within a glass tube in consecutive groups.* Either extremity of this apparatus will at any time affect an electrometer distinctly, without any previous preparation. Its power must consequently arise from the contact of the different materials of which it is composed. Now, as this is the most simple instance of electrical excitement; and as the

* A more ample description of the construction and properties of the electric column will be found in another part of this work.

effects produced are permanent, it is highly probable that previous contact may be the remote cause of the effects attendant on other processes.

However various the means employed to excite electricity, its effects are constantly the same; but certain phenomena observed at a very early period, shew that there is a distinction between the causes by which they are produced. Sealing-wax and glass, for instance, equally rubbed, will either of them occasion divergence in the leaves, or balls, of an electrometer, when presented to it separately; but if they are applied together, no effect is produced. Again, if the electrometer has been made to diverge by contact with excited wax, such divergence will be lessened by the approach, and destroyed by the contact of excited glass. Or if it be first electrified by excited glass, the electricity will disappear when excited wax is presented to it.* Here then appears *two* elec-

* Some care is necessary in these experiments, which can only become familiar by practice. The sealing-wax and glass should not be excited more powerfully than is sufficient to affect the electrometer distinctly; and when the leaves are intended to remain divergent, the excited body should be brought into actual contact with the cap of the instrument.

tric powers, similar in their *separate* action on the electrometer, and other indifferent matter; but exerting a mutual influence on each other, destructive of their individual properties.

It was at first conceived, that these phenomena were peculiar to the substances by which they were produced; and hence the power excited by rubbing glass was called vitreous electricity; and that resulting from the friction of sealing-wax, resinous electricity; but it is now demonstrated, that both powers are produced in every case of electrical excitation, and as their mutual counteraction of effect resembles that of an affirmative, and a negative power; the terms positive, and negative electricity, have been substituted for vitreous, and resinous.

The practical determination of these states in different excited bodies, is of importance to the electrician, and may be thus effected. Sealing-wax, when rubbed on woollen cloth, is negatively electrified. Glass, when rubbed by silk, is positively electrified. Let an electrometer be made to diverge by the contact of excited sealing-wax; whilst thus diverging approach it with any excited body, whose electricity is to be determined. If the divergence of the electrometer increases, the presented body is negative; if it is diminished, the presented body is positive. In other words, all those substances that lessen the divergence occasioned by excited wax, are positive; and such as increase it, negative: whilst those which lessen the divergence produced by excited glass, are negative; and such as increase it, positive.

If we examine by this test, the effects produced in some of the instances of excitation already considered, the truth of the preceding statements will appear, and the relation of the different electrical states to the processes by which they are produced, will become more intelligible.*

Experiment 1. Roll up a warm and dry flannel, so as to admit of its being held by one extremity, whilst a stick of sealing-wax is rubbed with its opposite end. After slight friction, present the flannel to an electrometer, which

* After every experiment, the divergence of the electrometer should be destroyed, unless it be otherwise stated. This is effected by touching its cap with the finger, or a piece of wire.

will diverge; whilst the divergence continues, bring the stick of sealing-wax near the cap, and the leaves of the electrometer immediately close. Sealing-wax and woollen cloth are therefore both electrified by mutual friction; but their electricities are opposite; the wax being negative, the woollen positive.

Experiment 2. The electrical powers thus excited, are equal to each other; for, if the friction be repeated, and the wax and flannel be both presented at once to the electrometer, no signs of electricity appear. The opposite electricities, when applied together, producing a reciprocal counteraction of effect.

Experiment 3. Excite a black and a white silk ribbon, in the manner described at the commencement of this chapter. On separation, the black ribbon will be found negative; the white one positive.

Experiment 4. Take the sulphur cone, (formed by pouring melted sulphur into a conical wine glass), apply the cone, and the glass, separately, in succession to the electrometer; the former will be found negative, the latter positive.

Experiment 5. Apply the opposite ends of

the electric column, alternately, to the electrometer; they will be found differently electrified: the end terminated by zinc, being positive; and that terminated by silver, negative.*

Hence it appears, that positive and negative electricity are produced at the *same time* in all our experiments, and may be observed when proper means are employed for that purpose. But it is also seen, that by friction with the same substance, different bodies are variously affected; for glass rubbed with silk evinces positive electricity; but sealing-wax rubbed with silk is rendered negative. Again, polished glass, when rubbed with silk, skin wool, or metal, becomes positive; but if it be excited by friction against the back of a living cat, it appears negative. Wool, silk, or fur, rubbed against sealing-wax, are rendered positive; but gold, silver, or tin, are by the same process rendered negative.

* Some difficulty attends the negative electrization of the gold leaf electrometer, by excited wax; which sometimes separates the leaves so powerfully, as to destroy them. The electric column is more convenient for this purpose. A short contact of its silver extremity with the cap of the instrument will communicate the proper negative divergence. It may therefore be employed in the preceding experiments.

Tables have been formed exhibiting these effects between a variety of substances. The following is given on the authority of Mr. Cavallo:—

	Is rendered	*By friction with*
The back of a cat	Positive	Every substance with which it has been hitherto tried.
Smooth Glass. .	Positive	Every substance hitherto tried, except the back of a cat.
Rough Glass. . . .	Positive	Dry oiled silk, sulphur, metals
	Negative	Woollen cloth, quills, wood, paper, sealing-wax, white-wax, the human hand.
Tourmalin.	Positive	Amber, blast of air from bellows.
	Negative	Diamonds, the human hand.
Hare's skin	Positive	Metals, silk, loadstone, leather, hand, paper, baked wood.
	Negative	Other finer furs.
White silk	Positive	Black silk, metals, black cloth.
	Negative	Paper, hand, hair, weasel's skin.
Black silk.	Positive	Sealing-wax.
	Negative	Hare s, weasel's, and ferret's skin, loadstone, brass, silver, iron, hand, white silk.
Sealing-wax. . . .	Positive	Some metals.*
	Negative	Hare's. weasel's, and ferret's skin, hand, leather, woollen-cloth, paper, some metals.
Baked wood. . . .	Positive	Silk.
	Negative	Flannel.

* Mr. Cavallo had inserted metals, which appeared to imply that the friction of all metals electrified sealing wax positively; this I find is not the case: iron, steel, plumbago, lead, and bis-

The result of experiments of this kind is much influenced by the state of the bodies employed, and the manner in which friction is applied to them. In general, strong electric signs can only be produced by the friction of dissimilar bodies; but similar substances, when rubbed together so that the motion they individually experience is unequal, are sometimes electrified; and, in such cases, the substance whose friction is limited to the least extent of surface, is usually negative. Such is the case with the strings of a violin, over a limited part of which the bow passes in its whole length, and the hairs of the bow become positive. Similar is the effect when two ribbons of equal surface are excited by drawing one lengthwise over a part of the other; that which has suffered friction in its whole length becomes positive, and the other negative.

muth, render sealing-wax negative, and all the other metals I have tried leave it positive. I have therefore made a slight alteration in the table. The least difference in the conditions of such experiments will occasion singular varieties of result; with the same rubber (an iron chain), positive electricity may be excited in one stick of sealing-wax and negative in another, if the former have its surface scratched and the latter be perfectly smooth. Many repetitions of each experiment are therefore essential to an accurate conclusion.

From these facts we learn that positive and negative electricity are concomitant phenomena, and that in *all* cases of electrical excitement, they are *both produced*, though one only may occasionally appear, (a circumstance whose cause will be soon explained). It is seen also that these phenomena are not peculiar to any distinct class of bodies, but may be produced indifferently, or alternately, in various substances, by changing the materials or method by which friction is communicated to them. This knowledge simplifies the appearance of such electrical effects as have been here considered, by referring them to a general origin; but the peculiarities of electrical action are not yet sufficiently developed to authorise any present speculation on its cause.

CHAP. II.

Of Conductors and Non-conductors of Electricity, and of the Electrical Apparatus.

It has been seen by the permanent divergence of the electrometer, when an excited electric is brought in contact with it, that electricity can be communicated or conveyed from one body to another; and a history of the discoveries of Mr. Grey, on this subject, has been given in the Introduction. But the faculty of electrical transmission is very different in different bodies; some convey it with great rapidity; others more slowly; and there are some that appear absolutely to arrest its progress. Examples of this fact are apparent in the most simple experiments. The divergence of an electrified electrometer may be destroyed, weakened, or maintained, by touching its cap with different bodies; now, as the divergence of the electrometer is caused by its electricity, such effects can only be produced by the relative power of the touching bodies to deprive it thereof; for whilst the electricity remains, its divergence will continue unaltered.

Experiment 6. Touch the cap of an electrified electrometer with a stick of dry glass, sulphur, or sealing-wax. The divergence of its leaves will continue. These substances then, do not transmit electricity.

Experiment 7. Touch the cap of the electrified electrometer with a piece of wood, a rod of any metal, a green leaf, or with the finger. Its divergence immediately ceases. Such bodies therefore permit the transmission of electricity.

By experiments of this kind it is found, that there is a gradation of effect from one class of bodies to the other. Those which transmit electricity with facility are called Conductors; those whose transmitting powers are inferior, Imperfect Conductors; and such as have no power of transmission, Non-conductors: but in general the various bodies in nature are divided into two classes only; the remote extremes of each forming the intermediate class.

In the following enumeration of the principal Conductors, and Non-conductors, the substances are placed nearly in the order of their perfection; but the determination of this circumstance has not hitherto been accomplished with much precision.

Conductors.

All the known metals.
Well-burnt charcoal.
Plumbago.
Concentrated acids.
Powdered charcoal.
Diluted acids, and saline fluids
Metallick ores.
Animal fluids.
Sea water, spring water.
River water, ice and snow.
Living vegetables.
Flame, smoke, steam.
Most saline substances.
Rarified air. Vapour of alcohol and ether.
Most earths and stones.

Many of the preceding substances fail to conduct electricity when they are made perfectly dry; hence it is concluded their conducting power arises from the water they contain. Indeed this faculty does not permanently exist in many of the bodies enumerated, but varies or disappears with their modifications of temperature, &c. Thus hot water is a much better conductor than cold water; and such is also the case with charcoal, and other substances.

Non-conductors.*

Shell-lac, amber, resins.
Sulphur, wax, jet.
Glass, and all vitrifications; talc.
The diamond, and all transparent gems.
Raw silk, bleached silk, dyed silk.
Wool, hair, feathers.
Dry paper, parchment, and leather.
Air, and all dry gases.
Baked wood, dry vegetable substances.
Porcelain, dry marble.
Some silicious and argillaceous stones.
Camphor, elastic gum, lycopodium.
Native carbonate of barytes.
Dry chalk, lime, phosphorus.
Ice at — 13° of Fahrenheit's thermometer.
Many transparent crystals, when perfectly dry.

* Nonconductors are also sometimes called electrics, and occasionally insulators; but the latter term is only applicable to the most perfect of them.

The ashes of animal and vegetable substances.

Oils; the heaviest appear the best.

Dry metallic oxides.

The most perfect non-conductors become conductors by the accession of moisture; hence the necessity of preserving them clean and dry during electrical experiments. Resinous substances, raw silk, and Muscovy talc, are least liable to attract moisture, and are therefore most useful where perfect non-conductors are required. Glass becomes moist only on its surface, and this tendency may be checked by covering the surface with sealing-wax or good varnish. Glass consequently enters most extensively into the structure of an electrical apparatus; its strength, and the facility with which it may be procured of any form, fitting it most admirably for that purpose.

Many substances in the preceding list lose their non-conducting power, and become conductors when intensely heated. Such is the case with red-hot glass, melted resin, wax, &c.; but the most intensely heated air, if unaccompanied by flame, is not a conductor. Many fibrous substances attract water so readily, that it is absolutely necessary to dry and warm them before their non-conducting property appears; this is particularly the case with paper, flannel, parchment, leather, &c. The influence of heat on this property is indeed very remarkable, and not perfectly intelligible; it is well exemplified in the following instance: Wood in its natural state is a conductor; if baked, its moisture is expelled, but its organization is not altered; it is then a non-conductor. By exposure to a greater heat its volatile elements are dissipated, and its indestructible base (charcoal replete with alkali) only remains; this is a conductor; but if exposed again to heat, with access of air, it suffers combustion, and is converted into ashes and gases, which are non-conductors.

There does not appear any definite relation between the chemical characters of bodies and their conducting powers; for the best conductors, (metals) and the best non-conductors, (resins, sulphur, &c.) are alike inflammable substances. The products of combustion too, are dissimilar in this respect: acids and alkalies conduct electricity, but the metallic oxides do not. Neither does it appear that specific gravity, hardness, tenacity, or crystalline arrangement of particles, are connected with the power of electrical transmission; for similar characters of this kind are possessed by bodies of both classes. Thus platina, the densest of bodies, is a conductor; but so also is charcoal and rarified air. Carbonate of barytes has great density, and is a non-conductor; but dry air, and the different gases, which are amongst the rarest forms of matter known, are of the same character. Many non-conductors are brittle; but some also are elastic, and others fluid; and there are bodies of all these classes that are conductors.

Whatever be the cause of nonconducting power, it is evident that without its existence as a property of air, and other substances, electrical phenomena would be unknown; for if the faculty of electrical transmission existed universally, the cause of every effect of this kind would be dissipated and lost at the moment of its production. But by the property of non-conductors any excited electricity which they surround is preserved; and it is then said to be *insulated.* A support of glass, sealing-wax, silk, or any nonconductor, is for the same reason called an *insulating support*, or an *insulator*; and a piece of metal or other conductor so supported, is named an *insulated conductor*.

The use of insulators and conductors in practical electricity may be exemplified by very simple experiments, which will form no improper introduction to the consideration of more important apparatus.

Experiment 8. Hold a sheet of writing paper before a fire till it is perfectly dry and warm; lay it flat upon a table and rub the upper surface briskly with Indian rubber. The paper will adhere to the table, and if lifted up by one corner and presented quickly to any flat conducting surface, as the wainscot, &c. will be attracted by and adhere to it. This adherence is occasioned by the attraction of electricity excited on the paper, which in its dry state is an insulator or nonconductor; the necessity of which circumstance to the success of the experiment is rendered evident by the paper falling down as soon as it has attracted moisture enough to destroy its insulating property, and is further apparent from the impossibility of producing the same results by the friction of paper in its ordinary state of dryness.

Experiment 9. Repeat the excitation of the

paper in a dark room; when the paper is lifted from the table by its corner, present the knuckle of the other hand successively to various parts of its surface, a series of faint divergent flashes of light will ensue. This light is occasioned by the transmission of the electricity excited on the paper to the hand; and it occurs at every contact, because the *nonconducting* power of the paper *prevents* its transmission from one part of the surface to another, the effect existing over the whole portion that has been subjected to friction.

Experiment 10. Excite the dry sheet of paper with Indian rubber, as before, and place it immediately on an insulating stand, Figure 3, consisting of a round plate of metal about six inches diameter, supported on a pillar of glass. Present the knuckle to the edge or under side of the metal plate, a bright spark will appear; but a second approach of the knuckle will produce either a very trivial effect, or none that is perceptible; for the metal is a *conductor*, and transmits the *whole effect* of the excited electric at once. Insulated conductors then are employed in the electrical apparatus to receive or collect the diffused electricity of excited bodies, and to apply it to the purpose of experiment.

The structure of an electrical apparatus consists in the judicious arrangement of insulators and conductors, so that the former shall prevent the dissipation of the effects the latter are employed to collect or transmit; thus the cap and leaves of the gold leaf electrometer form a conductor intended as a test of electrical action; but to fit this conductor for its purpose it is *insulated*, being supported on the glass cylinder by which the leaves are enclosed.

When electricity is excited by friction the quantity of effect is, within certain limits, proportioned to the extent of the rubbed surface; hence it appears that every part of that surface is concerned in the production of the general effect. Now, that this may be the case, it is essential that every part of such surface be insulating; for friction is a progressive process, a succession of contacts; and the effect produced by it in the first instant would otherwise be destroyed by conducting power, before a second operation could contribute to its increase. For this reason electricity is most usually excited by the friction of a conductor of limited

limited size, against the extensive surface of a nonconductor.

An apparatus properly arranged for the excitation of electricity is called an electrical machine. Usually, to excite positive electricity, a glass tube, about an inch in diameter and two feet long, is rubbed lengthwise by a piece of dry oiled silk held in the hand, which is made to grasp the tube. In this way both the silk and the tube are electrified; but the electricity of the silk is destroyed by the conducting power of the hand, and that of the tube only appears. In a similar way negative electricity is procured by rubbing a tolerably large stick of sealing-wax with dry flannel or fur; the electrical power of the sealing-wax being all that results. Thus with the most simple machinery two processes are employed to procure the opposite electricities, although they are both excited in each; but to obtain them both, it would be necessary to insulate the silk, or flannel, used as rubbers, either by employing them in a very dry state, rolled up, so as to produce the friction with one extremity, at a distance from the hand, or by affixing them to a glass or other nonconducting support. And neither of these methods would be convenient where many experiments are to be made. This difficulty does not occur when large surfaces of glass are employed instead of tubes as sources of excitation; for these may be made circular, and proper friction be communicated to them from a fixed cushion, on an elastic support, against which they are made to revolve. There are two forms of the electrical machine constructed on these principles, which have each peculiar advantages. The one was first proposed by Dr. Ingenhouz, and has been perfected by Mr. Cuthbertson. The other originated with the German electricians, and was greatly improved by Mr. Nairne.

Mr. Cuthbertson's machine consists of a circular plate of glass, turning on an axis that passes through its centre; it is rubbed by two pairs of cushions fixed at opposite points of its periphery by elastic frames of thin mahogany, which are made to press the glass plate between them with any required degree of force, by means of regulating screws. A brass conductor, supported by glass, is fixed to the frame of the machine, with its branched extremities opposite each other, and near the extreme diameter of the plate, in a direction at right angles to the vertical line of the opposite cushions. The branched extremities of the conductor are furnished with pointed wires, that serve to collect the electricity from the surface of the excited plate.—The machine is represented by Fig. 4.

Such machines have considerable power, and may be constructed on a scale of greater magnitude than those of any other form; they are therefore highly useful when great electric power is required; but they are seldom constructed so as to exhibit both electricities, because it is difficult to insulate the rubbers, and at the same time preserve the compact form of the machine.

The most simple and perfect machine is represented by Figure 5. It consists of a cylinder of glass, from 8 to 16 inches diameter, and from 12 to 24 inches long, turning between two upright pillars of glass, fixed to a stout mahogany base. Two smooth metal conductors equal to the length of the cylinder, and one third of its diameter, are placed parallel to it upon two similar glass pillars, which are cemented into two separate pieces of mahogany that slide across the diameter of the base, so as to keep the conductors parallel to the cylinder while they are brought nearer to, or placed further from, its surface at pleasure. One of the conductors has a cushion fastened to it by a bent metallic spring; the surface of this cushion is accurately fitted to the radius of the cylinder; it may be from eight to ten inches long, and from one inch and a quarter to one and three quarters wide. To the upper part of the cushion a flap of thin oiled silk should be attached; it is to be sewed on the face of the cushion about a quarter of an inch from its top edge, so that the silk at its juncture with the cushion may form a neat straight line, rising a little above the surface. The silk flap should reach from the cushion, over the upper surface of the glass cylinder, to within about an inch from a row of points that are attached to the side of the opposite conductor The conductor to which the cushion and its silk are fastened is called the negative conductor, because it exhibits the electricity of the cushion: the opposite conductor collects and displays the electric power of the glass cylinder, it is therefore called the positive conductor. Each conductor is perforated in various parts with holes about the size of a goose-quill, for the convenience of attaching wires and different articles of apparatus; and that which carries the cushion and flap has its sliding mahogany base attached to the bottom of the machine by an adjusting screw, that serves to regulate the pressure of the rubber against the glass. The motion of the cylinder is always in the direction of the silk flap; it may be communicated either by a simple handle, or by multiplying wheels; the latter produce more electricity in less time, but increase the labour of turning.

The facility with which the electric power of glass is excited varies with the nature of the surface employed as a rubber. Dry silk is very efficacious, but the most powerful effects are produced by the use of an amalgam of tin, zinc, and mercury, applied, by means of hog's-lard, to the surface of leather, or oiled silk. The cushion of an electrical machine is always coated on the side which performs the office of rubber with an amalgam of this kind, which should be spread evenly over its surface until level with the line formed by the seam which joins the silk flap to the face of the cushion. No amalgam should be placed over this seam, or on the silk flap, which last should be wiped clean whenever the continued motion of the machine shall have soiled it, by depositing dust or amalgam on its surface. The same attention is requisite to the surface of the glass, which often becomes covered with black spots and lines; more particularly when the amalgam has been recently applied, as they then appear in great abundance. These it is essential to remove as often as they are formed in any quantity, since they tend to lessen the power of the machine. The surface of the amalgamed cushion is also soon soiled; for the excited glass constantly attracts dust from surrounding bodies, and this dust is wiped off by the rubber as the glass passes it. If the dust is removed after every course of experiments, by separating the cushion from the negative conductor, and gently rubbing its surface and the surface of the silk flap with a dry linen cloth, the machine may be kept nearly in uniform good order, without a very frequent renewal of the amalgam; which is only necessary when that which has been applied becomes irregular on the surface of the cushion, or impregnated with dust from long use, or inattentive cleansing.*

The various articles of apparatus employed with the machine consists principally of insulating stands or supports, of various forms, and of wires and flexible conductors, by which a proper connexion with either of the conductors of the machine may be obtained. The application of the apparatus to the purpose of experiment will best explain the nature of its subordinate parts, and to this I shall now proceed.

* The amalgam I use is made by melting together one ounce of tin and two ounces of zinc, which are mixed, whilst fluid, with six ounces of mercury, and agitated in an iron or thick wooden box until cold. It is then reduced to very fine powder in a mortar, and mixed with sufficient hog's-lard to form a paste. Amalgams have sometimes a much larger proportion of mercury, but their action is more variable and transient; as is also the effect of their partial application to the surface of the machine during its action, as recommended by some electricians.

CHAP. III.

Experiments with the Electrical Machine.—Theory of its Action.—Phenomena of Attraction and Recession.

THE electrical machine being prepared agreeable to the directions in the preceding chapter, and the cushion pressed moderately against the glass by the action of its adjusting screw, it may be put in motion, and the following phenomena will be observed.

1st. Distinct lines of light, accompanied by lateral scintillations, pass from one conductor to the other across that part of the glass cylinder which is not covered by the silk flap. These are called electrical sparks.

2d. Bright sparks pass between either of the conductors and the knuckle, or any smooth uninsulated substance presented to them at a moderate distance; and if received on the knuckle or any part of the body produce a painful sensation.

3d. These effects are more distinct, and the sparks from each conductor stronger, when they are taken from both at the same time.

4th. The power of the spark from either the positive or negative conductor, singly, will reach its maximum when the opposite conductor is uninsulated, by suspending a chain or wire from it to the ground.

5th. If the two conductors are connected by a wire, or other conductor, the most vigorous friction of the cylinder will not electrify either.

6th. If, instead of a wire, the conductors are connected by a silk string on which a number of shot or metal beads are strung at the distance of a twentieth of an inch from each other, a series of bright sparks will pass between the beads as long as the turning of the machine is continued.

It is to be remembered, that the conductor to which the cushion is attached indicates the electrical phenomena of the cushion, and the opposite conductor exhibits the electricity of the glass cylinder; therefore the observation of their phenomena is virtually an observation of the circumstances that occur in all cases when electricity is excited by friction.

The first and second phenomena seem to shew that the cause of electricity is corporeal; for sensation is affected by it, and a mechanical impulse experienced, which it is difficult to ascribe to any other than a material cause.

The third phenomenon proves that there is a *mutual action* between the electricities excited in the *opposite* conductors; since their effects are *more powerful* when directed at the *same time* to *one* conducting body.

The fourth phenomenon shews that the *same* relation which is observed between the *opposite* electrified conductors exists also between *either* of them and the *ground*, but in a different degree.

By the fifth phenomenon it is seen that *positive* and *negative* electricity, if excited to the same extent, and *united* by *conducting matter*, exhibit no electrical phenomena.

The sixth phenomenon is observed merely to shew that when the conductors are *connected*, the machine continues to *excite* electricity, but is prevented from *displaying it* by their mutual contact.

From the consideration of these appearances the following explanation of electrical phenomena may be rationally deduced.

PROPOSITIONS.

1st. The *cause* of electrical phenomena is *material*, and possesses the properties of an elastic fluid.

2d. This *electric fluid* attracts and is attracted by all other matter, and, in consequence of such attraction, exists in all known substances.

3d. The attraction of different bodies for the electric fluid is *various*, and so is that of the *same body under different circumstances*; consequently the quantity of electricity naturally existing in different substances may be *unequal*; and the *same body* may attract *more*, or *less* than if *alone*, when *combined* with other matter: but its original attraction will be restored by destroying the artificial combination.

4th. From some peculiarity in the nature of the electric fluid, its attraction by and for common matter is more influenced by *figure* than by *mass*; and is consequently stronger in extensive than in limited surfaces.

5th. From the same peculiarity, the electric fluid moves with great facility over the surface or through the substance of some bodies, and is arrested in its progress by others.

6th. When the attraction of any substance for electricity is *equal* to the *electric fluid it contains*, that substance will evince no electrical signs; but these are immediately produced when there is either *more* or *less* electric fluid than is adequate to the *saturation* of the existing attraction: if there be *more*, the electrical signs will be *positive*; if *less*, they will be *negative.*

Electrical excitation then may be thus effected:—The bodies employed have *each* a *certain quantity* of the electric fluid proportioned to their natural attraction for it: this they retain, and appear unelectrified so long as they remain in their *natural* state. Now if two such bodies are brought in contact their natural attractions are *altered*, one of them *attracts more* than in its separate state, and the other *less*; the electric fluid diffuses itself amongst them in quantities proportioned to their *relative attractions*, and they consequently appear unelectrified. But if they are suddenly *separated*, the *new distribution* of the electric fluid remains, whilst the *original attractions* are restored, and as these are *not equal* to each other the bodies will appear electrical; that whose natural attraction was *increased* by contact, having received an *addition* to its quantity of electric fluid, will be *positively* electrified; and that whose attraction was *lessened*, having *lost* a portion, will be *negative.*

Take, as an instance, the electrical machine:

let the attraction of the cushion for the electric fluid be represented by 20, and that of a similar surface of glass by 30, the sum is 50. Bring the bodies in contact, their attractions alter; that of the glass becomes 40, and that of the cushion is reduced to 10; the sum of these is still 50: the natural electricity therefore, though unequally distributed, is still equal to the sum of the attractions, and does not appear; for the cause of its unequal distribution (the contact) is still active. Separate the glass from the cushion, its original attraction of 30 will now only operate, but it has acquired 40 of electricity by contact with the cushion; the glass is therefore positive with a force equal to 10. The cushion also will now have its original attraction of 20, but its electricity amounts only to 10: it is therefore negative with a force equal to 10. And here is seen the reason why positive and negative bodies act more powerfully on each other than on indifferent matter, for their mutual difference is often twice as great as their individual; since if the latter be 10, the former may be 20.

The effects now described continually recur during the revolutions of the cylinder, every part of which is successively brought in contact with the cushion, and passes forward with the electricity it thus progressively acquires. The silk flap may be considered as a continuation of the rubber, which, by partially maintaining the altered attraction of the glass, prevents the tendency of the acquired electricity to pass back into the cushion. The surface of the glass, where it passes from beneath the silk flap, has not this compensation; hence the acquired electricity is there uncombined, and has a tendency to diffuse itself amongst the surrounding bodies: the conductor, with its row of points, is the nearest reservoir, and into this it passes, and the conductor becomes thereby *positively electrified.* During this process the cushion and its attached conductor constantly furnish electricity to the glass, and *they* are consequently negative in the same degree; but they have only a *limited* surface, and a *certain quantity* of natural electricity, and, if perfectly *insulated,* (that is, surrounded by nonconductors) can furnish only a definite portion; but if they are connected with the ground, whose surface is comparatively *unlimited,* they operate upon an extensive store, to the supply of which there appears no assignable bound. It is for this reason that the electricity of either conductor *separately* is more apparent when the opposite one is *uninsulated.*

The excitation of electricity thus appears analogous to the evolution and absorption of heat: simple mechanical touch is rarely attended by any perceptible change of temperature, but such change is usual in cases of chemical combination. During the solution of many salts there is an absorption of heat; by the union of acids and water heat is evolved; and by the contact of certain acids with inflammable matter even light and ignition are produced. Such circumstances prove that some of the most active powers of nature exist around us at all times, latent in their natural states of combination, but rendered active by the slightest change.

Positive electricity has here been considered as the effect of a redundance of the electric fluid, and negative electricity as a deficiency: hence when sparks or other electric phenomena occur between two oppositely electrified conductors, it is supposed that such appearances are produced by the electric fluid passing from the positive to the negative, which motion is occasioned by a tendency to regain its natural state of distribution. And on the same principle the sparks and other effects that take place between an uninsulated and a positively electrified conductor are presumed to arise from the superabundant electric fluid passing from the electrified conductor to the ground; whilst those which occur between a negative conductor and the ground result from the passage of the electric fluid from the latter to the former.

There are certain appearances that demonstrate this direction of the electric fluid with tolerable accuracy; and, but for them, it must have been considered as only a probable supposition: for the motion of electricity is too rapid to admit of the detection of its course by the eye, unless indeed under very peculiar circumstances.*

Experiment 11. Present a pointed wire to any negatively electrified body,—a divergent pencil of light will evidently pass from the

* With very large and powerful electrical machines, sparks are sometimes procured of from ten to twenty inches long; and such sparks always appear to pass from the positive to the negative, or from the positive to the receiving ball.

point to the electrified surface. Present a similar point to any positive surface,—the point will be illuminated by a neat luminous star. The light in this experiment is unquestionably produced by the motion of the electric fluid: the point is to be considered as a pipe capable of emitting or receiving it, and the appearances correspond with the supposed course of the electricity: for the *negative* surface is stated to have a *deficiency*, and the point presented to it is illuminated by a *divergent* pencil, which indicates that the *cause* of that light moves *from* the point to the negative body. The *positive* surface is said to have an *excess* of electric fluid, and the point presented to it is merely illuminated by a globular *spot of light*, an appearance which may well be conceived to attend the *entrance* of a subtle fluid into it.

If the points are *connected with* the oppositely electrified bodies, their appearance is precisely the *reverse* of that which occurs when they are *presented to* them; and such should be the case if the preceding supposition be correct.

Experiment 12. Fig. 6. represents two hollow metal balls about three-fourths of an inch diameter, insulated on separate glass pillars by which they are supported at two inches apart; the upper part of each ball is indented to form a small cup in which a fragment of phosphorus is to be placed. A small candle or lamp has its flame situated mid-way between the balls; one of them is connected with the positive and the other with the negative conductor of the electrical machine by means of a wire. When the balls are electrified, the flame is agitated, and inclines to that which is negative; this it soon heats sufficiently to fire the phosphorus it contains; whilst the positive ball remains perfectly cold, and its phosphorus unmelted. If the connecting wires be now reversed, so that the ball which was negative shall become positive, and that which was positive be rendered negative, the phosphorus in the latter will soon take fire. So that *electricity passes from the positive to the negative*, and transmits with it the heat of any intervening ignited body.

Experiment 13. Take the transfer plate of an air-pump, and affix to its centre, by a wire of three inches long, a ball of an inch in diameter; connect a similar ball, by a sliding wire, to the top of a receiver, and place this over the transfer plate, so that the balls may be opposite to each other, and at the distance of about an inch,—see Fig. 7. Exhaust the receiver accurately by means of an air-pump, connect the pump plate by a wire with the negative conductor, and the upper wire and ball with the positive. When the machine is turned, a current of beautiful purple light will pass from the positive to the negative ball, on which it breaks and divides into a luminous atmosphere entirely surrounding the lower ball and stem, and conveying most strikingly, the idea of a fluid running over the surface of a resisting solid which it cannot enter with facility. No appearance of light occurs on the positive ball, but the straight luminous line that passes from it: but if it be rendered negative, and the lower ball positive, these effects are entirely reversed.

When this experiment is made with due care, it furnishes a most satisfactory ocular demonstration of the course of the electric fluid; and few who witness it under such circumstances can entertain any doubt on that subject.

Electrical phenomena then are produced by the motion of a naturally *diffused* fluid, which, by certain processes, may be *accumulated in* some bodies and *taken from* others; but, tending constantly to an *equilibrium*, will, if unobstructed, restore its original diffusion by passing *from* those that have a redundance *to any* that are deficient; or, if none of these are near, to such as have only their natural quantity.

Positive and negative are merely comparative terms, expressive of different variations from the natural state. There are two standards to which, under different circumstances, these states are referred. When the effect is measured by the divergence of pith-balls or other light bodies suspended in the atmosphere, the ambient air becomes the standard of plus and minus, these states being then only indicated by the balls in proportion as they actually differ from it; and, as the air is a nonconductor, it may be considered as *insulated.* But if any substance connected with the ground be presented to an insulated electrified body, then the ground becomes the standard by which the positive or negative divergence of that body is measured. Hence the standard is called a neutral point, and all bodies having only their natural quantities of electric fluid may be regarded as such; and, although actually unelectrified, are to be considered positive when compared with such as have less than their natural portion, and negative when opposed to those that have more.

The motion of light bodies produced by electricity, and usually called attraction and repulsion, is occasioned by the mutual attraction existing between the electric fluid and common matter. Its nature will be best understood by reference to experiment.

Experiment 14. Take a small downy feather, or a pith-ball suspended by a metal thread (such as is used for gold lace), and holding the thread, bring the ball near any electrified conductor, either positive or negative: the ball will be attracted by and adhere to the electrified conductor, and will remain in contact with it until its electricity is destroyed.

Such bodies as are *positively* electrified, tend to *diffuse* their superabundant fluid amongst surrounding substances; and those that are *negative*, endeavour to *acquire* electric fluid: hence either state of electricity will produce attraction; for if light bodies are to be moved, it is indifferent whether the electrified surface attracts their *natural electric fluid*, or the *matter* to which it is attached; for the attraction arises only from the *different proportions* of these in any *two bodies*, and will of course *continue* whilst that *difference exists*. Now, in the preceding experiment, the attracted body was in *conducting* communication with the ground, and the electrified surface, being comparatively of *limited* extent, could not perceptibly alter the electric state of the whole earth, the attraction must therefore continue until the electrified body has received from, or communicated to, the earth such a portion of the electric fluid as it is deficient in or overcharged with; and, consequently, till all *electrical difference* between the earth and it is *annihilated*.

Experiment 15. Repeat the preceding experiment with a ball or feather supported by a *silk* thread: the light body will first be attracted to the electrified conductor, and will then *recede* from it; nor can again be brought in contact until it has touched some uninsulated conducting substance.

The light body is here attracted for the same reason as before, but it is *insulated*, and consequently receives, by contact with the electrified surface, a *similar* electric state; it therefore recedes from that surface, being *attracted* by the *ambient air*, or other surrounding bodies, for *they* have their *natural portion* of electricity, and therefore *differ* from the light body, which has either *more* or *less;* but the electrified surface does not differ from the light body, and, consequently, cannot attract it, till, by touching some uninsulated conductor, its natural electric state is restored.

From these experiments it necessarily follows,—

1st. That bodies *positively* electrified in the *same degree* will recede from each other; their *plus electric fluid* being attracted by the ambient medium, which is in its *natural state.*

2d. *Negative* bodies of *equal* power will recede from each other, their *matter* having an attraction for the *natural electricity* of the surrounding medium.

3d. Bodies electrified either positively or negatively, in *different* degrees, will be *mutually* attracted, until their relative proportions are *equalized,* when they will *recede* from each other and tend towards unelectrified substances.

4th. Positive and negative bodies will reciprocally attract each other, and, if of *equal intensity,* be unelectrified by contact.

5th. Such as are positive and negative in *different degrees* will attract each other, and remain electrical, after contact, in proportion as the *sum* of their electricities may *deviate* from the state of the surrounding medium.

Hence we may conclude, that when any two substances *recede* from each other, they are *similarly* electrified; and when they *attract* each other, they are *oppositely* electrified.

By the operation of these principles a variety of entertaining experiments may be made; for light substances placed between differently electrified conductors will move from one to the other, and by such alternate motion produce some singular results.

In the following experiments, which are common illustrations of electric motion, the moving body is always situated between an electrified surface and one that is in communication with the ground: it is first attracted by the electrified surface, because it is in a *different* state; by contact its electricity becomes the *same*, and it is then attracted by the body in connection with the ground; it touches that, has its natural electricity restored, and is then reattracted by the electrified surface, becomes again electrical, and recedes to its original situation, whence it is again attracted, &c. and this motion must necessarily continue until the *different* electrical states of the *two* surfaces are *equalized* and rendered *similar*. Light substances moved by electricity may therefore be considered as vehicles of transfer, conveying the electric fluid from one system of bodies to another, and thus promoting its natural distribution.

Experiment 16. Place a leaf of gold, silver, or Dutch metal on the palm of the hand, and bring it within a few inches of an electrified conductor; it will be attracted and continue to move, alternately from the hand to the conductor, as long as the latter is electrified.

Experiment 17. Suspend a brass plate from the conductor of an electrical machine, and beneath it, at the distance of three or four inches, place a similar brass plate connected with the ground, on this put some small figures cut in paper; when the upper plate is electrified the figures will rise and perform an electrical dance by their motion between the plates.

Experiment 18. Place a pointed wire on the machine, electrify the inside of a dry glass tumbler by holding it over the wire whilst the machine is in motion; place some pith balls on the table and cover them with the electrified glass; they will be alternately attracted by it and the table, and continue their motion for some time.

Experiment 19. Insulate two small bells on separate glass pillars, at three-fourths of an inch distance from each other; suspend a clapper by a silk thread so as to hang midway between them; connect one of the bells with the conductor of the machine and the other with the ground; the clapper will vibrate from one to the other during the action of the machine, producing an electric chime.

Experiment 20. Insulate a circular ring of brass so as to stand near an inch and a half from the flat surface of a table; connect the brass ring with the conductor of the electrical machine, and place within it, on the table, a very light and round glass ball of two inches diameter; the ball will be attracted by the ring, touch it, and become electrified at the point of contact; this point will then recede and be attracted by the table, whilst another part of the ball is attracted by the ring; and, by the repetition of this process, the ball is made to revolve and travel round the circumference of the ring. This phenomenon depends on the nonconducting power of the ball, which confines the effect of contact to a limited portion of its surface, different parts of which are consequently variously electrified at the same time.

These phenomena of electrical motion, which are all referred to the same principle, certainly evince the materiality of the electric fluid, which here, by its attraction, displays one of the most essential properties of matter, and even counteracts the effects of gravitation. The separation of the parts of similarly electrified bodies confirms the preceding evidence, as will appear by the following illustrations.

Experiment 21. Connect a pith-ball electrometer with each conductor of an electrical machine, both remaining insulated; when the machine is turned, each electrometer will diverge, for they both differ from the surrounding air (one having more electric fluid and the other less); connect the opposite conductors by a wire, the divergence will cease, for the electrical difference of the conductors and the ambient air is destroyed.

Experiment 22. Take a dozen threads and tie them together at top and bottom; annex them (by a loop attached to the upper knot) to the conductor of the electrical machine; when electrified the threads will separate from each other, and the knot at the bottom rising they will assume a spheroidal figure, which will continue as long as they are electrified.

Experiment 23. Insulate a condensed air-fountain, and electrify it; the jet will be minutely subdivided and expanded over a considerable space, but will return to its original limit when the electrization is discontinued.

Experiment 24. Fasten a piece of sealing-wax to a wire, and insert this in one of the holes in the conductor of the electrical machine; soften the sealing-wax by heat, and whilst it is still soft turn the cylinder; very fine threads of wax will be separated, and if received on a sheet of paper will cover it with minute fibres like fine red wool.

Experiment 25. It will be shewn hereafter that pointed bodies transmit electricity with greater facility than such as have blunt or rounded terminations; hence, if any electrified conductor have points on its surface, the air opposite those points is soon similarly electrified, recedes from them, and is replaced by other unelectrified particles, which also become electrical and recede; so that a current of air is constantly produced by an electrified point, and appears to issue from it, whether the point be positive or negative. On this principle various revolving motions are produced. Fig. 8. represents a wire cross with pointed extremities bent in one direction; when this is balanced by its centre on a point, and electrified, it turns swiftly round in the contrary direction to its points; the reaction of the air against the currents they produce being the cause of its motion.

Light models fitted up with vanes, like the floats of a water-wheel, may be put in motion by the current of air produced by the action of an electrified point; and if a lighted candle be presented to such a current, its flame will sometimes be blown out.

Such are the principal phenomena of motion produced by the action of electricity; they are susceptible of almost unlimited variety, but uniformly result from the simple principles already stated, namely, the attraction of the electric fluid for common matter; its tendency to equal diffusion; and the occasional interruption of these properties by nonconducting power and altered force of attraction.

CHAP. IV.

On the Phenomena of Electric Light.

THE luminous appearances produced by electricity exhibit considerable diversity; it is therefore necessary to consider them with attention, and compare the circumstances of their production with the general principles of electrical action.

Light is not constantly attendant on the excitation of electricity, but appears when that process is vigorously performed, and is then brilliant in proportion to the *intensity* of the excited electricity.

Suppose 10 particles of electricity to be added to or subtracted from a body whose natural attraction is for 25; the electrical *difference* between that body and the substances (in their natural state) by which it is surrounded will be 10; the *intensity* of its positive or negative state may be then expressed by 10. If the alteration in its natural quantity be now made equal to 20 particles, its electrical difference will be *twice as great*, and it will therefore have *double* the *intensity;* so that this term in electricity is employed to express the greater or less deviation of any electrified body from the standards of plus and minus.*

The light evolved in ordinary cases of excitation extends only to faint flashes and scintillations, sparks being only produced when these effects are concentrated, as they are in the electrical machine by the action of its conductors. The spark in passing from one body to another is influenced by the *form* of the conductors, their *extent*, and the *nature* and *density* of the medium through which it passes: it will be necessary to consider each of these separately.

The distribution of electricity on conductors has evidently little relation to their solid con-

* The elasticity of the electric fluid admits the arrangement of *more* or *less* particles in the *same space*, and its *intensity* or tendency to an equilibrium will be proportioned to the *quantity* accumulated in any *given surface*, or to its *density*: this corresponds with the action of other elastic fluids; air, for instance, of such density as to support a column of one inch of mercury, will sustain two inches when *compressed* into *half* its original space, and only half an inch when *expanded* to twice its original bulk, and will unite with water or other liquids in quantities proportioned to its density. Positive and negative electricity are analogous to condensed and rarified air.

tents, but depends almost entirely on surface, for the same effects are produced by the thinnest cylinder or sphere of metal as by the most compact solid body of the same form and dimensions; it is indeed even probable that the action of insulated conductors consists in the ready communication of their electric state to the contiguous surface of the extensive stratum of air by which they are surrounded, and to the facility they present to the discharge of that electrified stratum when an uninsulated or differently electrified body is brought near them; for every positively electrified conductor is surrounded by a positive atmosphere, and every negative conductor with a negative atmosphere whose densities decrease as the square of their increased distance. Hence any insulated electrified body will retain its electrical state until its intensity is sufficient to overcome the resistance of the air (which is the medium by which it may always be considered as separated from uninsulated or differently electrified bodies), and the greater or less interval through which the spark passes is called the *striking distance*.

When the surface of the conductor is uniform, the reaction of the air around it is also uniform; but if the surface of the conductor be irregular, the tendency of the electric fluid to escape or enter will be greatest at the most prominent parts, and most of all when these are angular or pointed. To understand this it is only necessary to recollect that every electrified conductor is surrounded by an atmosphere of its own figure, the contiguous surface of which is similarly electrified: and that electricity is *not* transmitted through air, but by the *motion* of its particles. For this motion of particles is *resisted* by a uniform surface from the *similar* action of the air around it, which is all *equally* capable of receiving electricity, and cannot tend to distribute it in one direction more than another; the immediate electrical atmosphere of the conductor will be therefore resisted in any attempt to recede from it by a column of air which is *equally* opposed in every part; but if there be any *prominent* point on the conductor projecting into the atmosphere, it will facilitate the recession of the electrified particles opposite to it by removing them *further from* the electrified surface, and *opposing* them to a *greater number* of such as are *unelectrified.*

The action of pointed or angular bodies consists then in promoting the recession of the particles of electrified air, by protruding a part of the electrical atmosphere of the conductor into a situation more exposed to the action of the ambient unelectrified medium, and thereby producing a current of air from the electrified point to the nearest uninsulating body. Hence the most *prominent* and the most *pointed* bodies are such as transmit electricity with the greatest facility, for with them this condition is most perfectly obtained.

A spherical surface is that which, considered with regard to its surrounding atmosphere, is most uniform; balls, therefore, or cylinders with rounded ends, are usually employed for insulated conductors, and their magnitude is proportioned to the intensity of the electrical state they are intended to retain; for a point is virtually a ball of indefinite diameter, and will indeed act as such with regard to very small quantities of electricity, and a ball of moderate size may also be made to act as a point by electrifying it strongly.

If two spheres of equal size are connected together by a long wire and electrified, their atmospheres will extend to the same distance, and they will of course have respectively the same intensity; but if the spheres be of unequal size, the atmosphere of the smallest will extend furthest, and it will necessarily have the greatest intensity; so that a longer spark can be drawn from a small ball annexed to the side of a conductor than from the conductor itself, and longer in proportion as the ball projects further from the side.* Hence the finer the point, and the more freely it projects beyond any part of the conductor to which it is annexed, the more rapidly will it receive or transmit electricity. M. Achard found that a single pointed wire, screwed in the centre of a circular piece of brass one inch and a half diameter, produced a greater effect in transmitting or receiving

* Mr. Cavendish,† Coulomb,‡ Laplace, and Poisson,§ have investigated the ratio of electrical intensity on the surface of different conductors; each analysis involves an hypothesis, but that of Mr. Cavendish appears to me most rational, although several circumstances exist that preclude an accurate experimental demonstration: it is indeed probable that the intensities are in the inverse ratio of the surfaces; proceeding from a flat surface where it is least, to a point where it may be considered as infinite.

† Phil. Trans. vol. lxi. p. 624. &c.

‡ Acad. des Sciences, 1786, 1787, 1788, 1789.

§ Memoires de l'Institut, 1812.

electricity than nine similar points screwed into the same base, the proximity of the nine points occasioning them to act nearly as one conducting surface of the same area.

By inserting a fine point in the axis of a large brass ball, from beneath the surface of which it may be protruded more or less by the action of a fine screw, the effect of a ball of any size may be obtained; when beneath the surface of the ball the point does not act, but in proportion as it is protruded it increases the transmitting power, and, if projected far enough, at length entirely overcomes the influence of the ball.

From the probable law of electrical distribution, stated in the preceding note, it follows, that the larger any insulated conductor may be, the greater will be the electrical charge it requires to pass through any given striking distance: hence very different effects are produced with the same electrical machine, when the size of its conductor is varied; and hence also sparks of the *same length*, taken from *different sized* conductors, must vary in *force*, as they do in *quantity*, of electric fluid. Very long and extended conductors give shorter sparks than such as are more compact, but they are sometimes more powerful.*

The following are illustrations of the influence of the *form* and *extent* of the conductor on the appearance of transmitted electricity.

Experiment 26. Present a brass ball of three inches diameter to the positive conductor of a powerful electrical machine; sparks of brilliant white light will pass between them, accompanied by a loud snapping noise: to produce these sparks in rapid succession the ball must be brought near the conductor, and they then appear perfectly straight.

Experiment 27. Annex a ball of an inch and a half or two inches diameter to the conductor, so as to project three or four inches from it; present the large ball to this, and much longer sparks will be obtained than from the conductor itself, but they will be less brilliant and of a zigzag form.

* Mr. Brook, of Norwich, formed a most extensive conductor of many long metal rods, suspended by glass sticks from his ceiling, and connected together nearly in the form of a grid-iron: with this apparatus the sparks, though shorter, were much more painful than those from a conductor of five feet long and five inches diameter, charged by the same cylinder.

Experiment 28. Substitute a small ball for that attached to the conductor in the former experiment; the electric fluid will now pass to a greater distance, but in the form of a divided brush of rays, but faintly luminous, and producing little noise: this brush will even occur with larger balls, if the machine be very powerful; it is most perfect when procured by presenting a flat imperfect conductor, as a piece of wood or paper, or the crown of a hat, instead of the large ball.

If, instead of a ball, a sharp point be affixed to the conductor, no sparks can be procured from it, but a divided brush of rays more minute than that in the preceding experiment will appear; and the electricity will be transmitted to a greater distance.

If the uninsulated body on which the sparks are received have its surface varied, the same diversity of result is obtained as by changing the surface of the conductor.

Experiment 29. Whilst a current of sparks are passing between a large ball and the conductor at the distance of an inch and a half, present a sharp point at twice that distance, and the sparks will immediately cease, the electric fluid being silently transmitted by the point.

Experiment 30. Enolose a point in a glass tube so that it may be placed at any distance from one of the open ends of the tube; in this situation its influence as a point will be destroyed, and it will transmit electricity by sparks as a ball. The power of a point is also destroyed by placing it between two balls, or in any way preventing its free and prominent exposure.

Experiment 31. Insulate a smooth metal cup with rounded edges, and in the cup place a quantity of smooth chain, free from sharp edges or points; let a silk thread be attached to one end of the chain, and passed over a pulley on the ceiling, so that the chain may be raised out of the cup at pleasure; attach a pith-ball electrometer to the cup and electrify it; raise the chain from the cup, and as it rises the divergence of the electrometer will diminish; lower the chain, and the original divergence will be restored. The cup and the chain form together a *conductor*, whose surface is *increased* by raising the chain, and this increased surface *diminishes* the *intensity* of its electricity by presenting it to a more *extensive surface* of unelectrified air.

If this experiment be made with a cup and chain of sufficient magnitude, well insulated, sparks may be employed as the test, and they will be most powerful when the surface is least extensive.

Electrical sparks are more brilliant in proportion as the substances between which they occur are better conductors; hence metals are almost exclusively employed for this purpose, wood and other imperfect conductors producing only faint red streams; yet these substances act as points with some efficacy, and particles of dust which collect around the apparatus are often troublesome to electricians from the same cause.

Electricity is not less affected by the nature and density of the medium through which it passes, than by the extent and figure of the transmitting conductors; usually its brilliance and force are proportioned to the density and non-conducting power of the medium in which it occurs; and hence it has been conjectured by Morgan,* and by Biot,† that light is extricated from those mediums during the rapid passage of the electric fluid by its mechanical compres-

* Phil. Trans. vol. lxxv. p. 198.
† Annales de Chimie. vol. liii. p. 321.

sion of their particles: an idea well supported by most of the experiments yet made on this subject.

The various forms of the spark proceed from the different modifications of the powers by which it is produced; namely, the velocity and quantity of the electric fluid, and the density and insulating power of the ambient medium: in the open air long sparks are always crooked, for the electric fluid moving with great rapidity condenses the air before it, and is then resisted in that direction more than laterally; it changes its course, condenses the air in a new direction, is resisted, and again turned aside; and this alternate deflexion produces the zigzag appearance. When the interval is short the spark is usually straight, or slightly curved, but its appearance is irregular; sometimes broken or interrupted in different parts, and mostly redder and less brilliant in the middle than at the extremities. It is probable that these irregularities arise principally from the heterogeneous nature of the atmosphere, for in a vacuum the short sparks are uniform, and the long ones rarely deflected.

For experiments on the influence of different

gaseous mediums a simple apparatus is required: a globe of glass about four inches diameter, having two necks capped with brass; to one of the necks a stop-cock is screwed, with a wire and ball projecting into the globe, another ball is attached to a wire that slides through a collar of leathers screwed to the opposite cap, so that the balls may be set at any required distance from each other within the globe.—See Fig. 9. This apparatus may be exhausted of air by connecting the stop-cock with an air-pump, and different gases may be thus introduced into it, or the air it contains may be rarefied or condensed, and the effect of these processes on the form of the spark examined. In condensed air the light is white and brilliant; in rarefied air, divided and faint; and in highly rarefied air, of a dilute red or purple colour. The effect of gases appears to be proportioned to their density; in carbonic acid gas the spark is white and vivid, in hydrogen gas it is red and faint.

In proportion as the rarity of any medium is increased, a less intensity of electricity is required to render it luminous: this fact may be illustrated by a very simple apparatus.

Experiment 32. Seal a short iron or platina

wire within one extremity of a glass tube of 30 inches long, so that the wire may project a little within its cavity, and screw a ball on the external end of the wire; fill the tube with quicksilver, and invert it in a bason of the same; a vacuum will be formed in the upper part of the tube, which will occupy most space when the tube is vertical, and gradually diminish as it is inclined; a spark which in the open air would pass through only a quarter of an inch, will pervade six inches of this vacuum with facility; and if the quicksilver be connected with the ground, a current of faint light will pass through the upper part of the tube whenever its ball is brought near an electrified conductor. If, previous to the inversion of the tube, a drop of water or of ether be placed on the mercury at the open end, and secured by the finger whilst the tube is inverted, it will rise to the top, and when the finger is removed and the quicksilver descends, the ether or water will expand and extend the vacuum, and through this expanded vapour a current of electricity will become luminous, and of various colours in proportion to its intensity; when the spark is strong, and passes through some inches of the expanded ether, it is usually of a beautiful green colour.

Experiment 33. Take an air-pump receiver of 12 or 14 inches high and 6 or 7 inches diameter; adapt a wire (pointed at its lower extremity) to the top of the receiver, letting the point project an inch or two into its inside; place the receiver on the plate of the air-pump, and electrify the wire at its top positively; whilst the air remains in the receiver, a brush of light of very limited size only will be seen, but in proportion as the air is withdrawn by the action of the pump this brush will enlarge, varying its appearance and becoming more diffused as the air becomes more rarefied; until at length the whole of the receiver is pervaded by a beautiful blush of light, changing its colour with the intensity of the transmitted electricity; and producing an effect which (with an air-pump of considerable power) is pleasing in the highest degree.

Even good conductors of electricity are rendered luminous by its passage through them, if they have sufficient tenuity.

Experiment 34. Insulate a large brass ball

and connect with it a silver thread of two or three yards long, the other extremity of which is held in the hand; when sparks are made to strike on the brass ball, the whole of the thread will be rendered faintly luminous.

The electric spark, viewed through a prism, exhibits all the prismatic colours, and is analogous to solar light in its power of displaying them separately by the intervention of different media; this is well exemplified by a very simple arrangement.

Experiment 35. Take a piece of soft deal, about three inches long and an inch and a half square; insert two pointed wires obliquely into its surface at an inch and a half distance from each other, and to the depth of an eighth of an inch; the wires should incline in opposite directions, and the track between the points be in that of the fibres; a spark in passing from one point to another through the wood will assume different colours in proportion as it passes more or less below the surface; and by inserting one point deeper than the other, so that the spark may pass obliquely through different depths, all the colours may be made to appear at once.

Sparks taken through balls of wood or ivory

appear of a crimson colour; those from the surface of silvered leather are bright green; a long spark taken over powdered charcoal is yellow; and the sparks from imperfect conductors have a purple hue. The quantity of air through which these sparks are viewed also influences their appearance; for the green spark in the vapour of ether appears white when the eye is placed close to the tube, and reddish when it is viewed from a considerable distance.*

Metallic conductors, if of sufficient size, transmit electricity without any luminous appearance, provided they are perfectly continuous; but if they are separated in the slightest degree, a spark will occur at every separation: on this principle various devices are formed, by pasting a narrow band of tin-foil on glass in the required form, and cutting it across with a pen-knife where sparks are wanted to appear, (the pen-knife should be passed twice over the strip of tin in opposite directions, as if to form the letter X, which will take out two conical pieces, and leave a small and well-defined separation). If an interrupted conductor of this kind be pasted round a glass tube in a spiral

* Morgan's Lectures, p. 234.

direction, and one end of the tube be held in the hand and the other presented to an electrified conductor, a brilliant line of light surrounds the tube, which has been hence called the spiral tube, or diamond necklace.—See Figure 10. By enclosing the spiral tube in a larger cylinder of coloured glass, the saphire, topaz, emerald, and other gems may be imitated. Words, flowers, and other complicated forms are also procured nearly in the same manner, by a proper disposition of an interrupted line of metal on a flat piece of glass, as may be seen in Figure 11. Indeed the tendency of the electric fluid to evolve light when it passes from one conductor to another is such that even their apparent contact does not entirely prevent it, for a chain will become luminous at every juncture of the links when an intense spark is passed through it. The light evolved by the action of a powerful electrical machine is so considerable, when a current of sparks are taken between two large balls, as to enlighten the whole of a very large room, so that the objects it contains may be distinctly perceived; and with a cylinder machine of 14 inches diameter, I have occasionally illuminated nine feet of spiral tube, in which the electric fluid became luminous at near eight hundred distinct separations.

Such are the principal phenomena of electric light; they are certainly conformable to the idea that it results from the mechanical action of the electric fluid on resisting mediums; but whether the light evolved is to be considered as a component part of those mediums, or of the electric fluid itself, these facts afford no data to determine.

CHAP. V.

On the Leyden Jar, and the Nature of Electrical Influence.

THE sources of electrical accumulation yet described consist of excited bodies and insulated conductors; and in the preceding chapter it has been shewn, that the form and arrangement of these last influence very materially the appearance of the electricity they convey.— When an electrified conductor has its surface *extended*, its *intensity* is *diminished;* and as this extension is virtually an exposure to a greater surface of *unelectrified air*, it might be expected that a similar effect would be produced by approximating the conductor to the ground; or to any other body of sufficient magnitude, in its natural electric state; and such is really the case.

Experiment 36. Insulate a flat metal plate with smooth rounded edges, and connect with it a pith-ball electrometer; electrify the plate either positively or negatively, and the balls will diverge: bring a similar plate *uninsulated* near that which is electrified, keeping their flat surfaces parallel and opposite to each other; the balls of the electrometer gradually collapse as the plates approach, and when they are within about half an inch of each other, the insulated plate appears unelectrified; but on the removal of the *uninsulated* plate the original divergence is restored.—See Figure 12.

When the insulated conductor is electrified, its pith-balls separate, because they are in a *different* electrical state to the air by which they are surrounded, whose matter or electric fluid they attract; but all unelectrified bodies have the *same relation* to the electrified balls as the ambient air has, and such as are *conductors* and connected with the ground present a more ample source of matter and electric fluid; consequently, if any such bodies are brought *near* the electrified conductor, its attraction is exerted on them, and the influence of the surrounding air is proportionably diminished; and if the proximity be sufficient, the attraction of the electrified surface will be so exclusively exerted in that direction as to be imperceptible in any other.

In this experiment the bodies are not brought

in *contact*, but only *near* each other, and consequently there is no *communication* or *loss* of electricity, but merely a *compensation* of its attractive power; hence when the uninsulated plate is *removed* the divergence of the electrometer is restored.

This fact shewing the *diminished intensity* of insulated electrified bodies when *opposed to uninsulated* conductors, supplies a method of increasing the positive or negative states (usually obtained) to a very great extent; for it is evident the electrical state of any body may be altered in the greatest degree whilst it is opposed to a conductor communicating with the ground, since, in that situation, its electricity will be compensated by the proximity of an exhaustless store, and cannot so soon acquire an intensity which would oppose its further progress, as the diminished or increased elasticity of the air by rarefaction or condensation limits the operations of the air-pump and condenser.

Since electrical attraction diminishes as the square of the increased distance at which it operates, the action of this principle of compensation will be *greater* in proportion as the *distance* of the opposed surfaces is *less*, provided a resisting medium be placed between them, to prevent the transmission of the electric fluid from one to the other. When air is the intervening body, it will resist only the passage of small quantities of electricity, as the mobility of its particles occasions it to yield to a very slight force: glass and Muscovy talc are the most compact of the solid nonconductors that can be reduced to their lamina, and these therefore form very fit media for the purpose of such experiments.

The insulating faculty of nonconductors depends on their impermeability to the electric fluid; and, in a perfect state, the most compact of them are never penetrated by it without mechanical injury: but electrical attraction is exerted through thin sheets of glass or other nonconducting matter with some facility: this has been considered by some as anomalous; but it is not more surprising than that the sun should act on bodies at many hundred millions of miles distance without any apparent medium of connexion: nor is it more remarkable that electricity and matter should act on each other through media that resist their transition, than that a magnet and a piece of iron should exert their mutual attraction when separated by bodies through which neither the magnet nor iron can pass.

If a sheet of glass then be placed between two plates of metal, one of which is connected with the ground and the other not; the insulated plate will have a greater capacity for electrical change than if *freely* insulated, and may be electrified either positively or negatively to a greater extent. The opposition of an insulated to an uninsulated conductor is the condition of the experiment; its success therefore does not depend on the *form* of the glass, but its *thickness*, which forms the medium of separation between the *metal plates;* they may be laid on the opposite surfaces of a glass plate, a sphere of the same substance, or a jar; but, in either case, the glass must extend two or three inches beyond the limit of the metal coatings, that they may be separated by a sufficient interval of air. The middle only of a plate of glass should be covered with metal, leaving an interval of two inches all round.—See Figure 13. A plate so prepared is called a coated pane.

The most convenient form is that of a cylindrical jar, covered on the inside and outside with tinfoil to within two or three inches of the top edge, the uncoated part must be kept clean and dry; to the inside coating a wire and ball should be attached, and rise two or three inches above the top of the jar.—See Figure 14. This is nearly the form in which the experiment was first made in the university of Leyden, it is therefore called a Leyden jar, or phial, and occasionally also an electric jar.

The metallic covering on the inner surface of the jar, with its attached ball and wire, is called the inner coating; the metallic covering on the outside of the jar, the outer coating; and the uncovered part of the rim, the uncoated interval.

Experiment 37. Present the knob of a Leyden jar to the conductor of an electrical machine at the distance of about half an inch, the jar being held in the hand by its outer coating; a series of sparks will pass to the ball of the jar, but will gradually grow weaker and at last cease. Remove the jar from the conductor and (its coating being still held by one hand) touch the ball of the jar with the other; a smart snap will be heard, and a violent and painful sensation be experienced, principally at the wrists, elbows, and across the breast. This singular sensation, which must be felt to be conceived, is called an electric shock: it is painful only for the moment, and leaves no permanent impression but that arising from surprise or fear.

Experiment 38. Make the uncoated interval very clean and dry, and place the knob of the jar in contact with the conductor, holding the jar by its outer coating as before; after a few turns of the machine flashes of light will be seen on the uncoated interval, and these will be soon followed by a loud explosion and a most brilliant electric spark, passing from one coating to the other: if you now touch the knob, but a very slight sensation will be felt, the electrical equilibrium being restored by the explosion and spark, which having occurred by the force of the accumulated electricity is called a *spontaneous explosion*, or discharge.

The application of the jar to the machine, to prepare it for the production of these effects, is called charging the jar; and any process of explosion by which the equilibrium is restored, its discharge. To effect a discharge, a communication must be made between the inner and outer coatings by some conductor, so that the power of the jar arises from the different electrical states of these coatings; and as their communication destroys all signs of electricity, they must be respectively positive and negative in an equal degree.

To avoid the shock during experiments with the jar, its discharge is usually effected by two knobbed wires, connected by a joint like a pair of compasses, and mounted on a glass handle: such an apparatus is called a discharging rod.—Figure 15.

To ascertain the degree in which the jar is charged, a particular contrivance, called Henley's electrometer (from the name of its inventor) is employed: it consists of a smooth round stem about seven inches long, with a ball on its top; immediately under this ball a semicircle of ivory is fastened to the side of the stem; in the centre of the semicircle a pin is fixed, on which a thin piece of cane, four inches long, with a light ball at its lower end, turns freely, and traverses the semicircle as an index. The lower half of the ivory semicircle is divided at the edge into 90 degrees. When this instrument is not electrified, its index hangs parallel to the stem; but when electrified, the light ball recedes and carries the index over the graduated circle to a greater or less extent in proportion to the intensity of the electricity.—See Figure 16. The recession of the index from the stem is greatest when it stands at right angles to it, or points horizontally; and having then moved over a quarter of a circle, it is said to indicate an electricity of 90 degrees. When the index is parallel with the stem, it stands at 0, or the commencement of the scale; and, as it gradually recedes, passes through 10, 20, 30, &c. to its horizontal position of 90. Hence this instrument is also called a quadrant electrometer.

The power of the Leyden jar is not more than proportioned to the time required to charge it: many hundred sparks will pass between the conductor and its knob during that process, and these are all concentrated into one spark when the jar is discharged; and hence the increased loudness of the explosion, brilliance of light, and acuteness of the sensation it produces.—The jar, when applied to the conductor of the machine, diminishes its intensity, and admits a much greater change in its electric state before any given resistance is overcome, or force of attraction manifested: this is seen by placing the quadrant electrometer on the conductor: if the machine is turned, the electrometer immediately rises to its limit, but when the knob of the jar is placed in contact with the conductor, the machine must be turned many times before this is effected, and the rise of the electrometer is then very gradual.

Experiment 39. The power of the jar, as a source of electric accumulation, depends on the opposite states of its two surfaces, which cannot obtain unless one of them is connected with the ground. Suspend a globular jar* by its knob from the positive conductor of the machine, its outer coating being surrounded by dry air cannot part with any of its natural electricity, and consequently the action of the most powerful machine will not communicate any charge to the jar (as will be seen by applying the discharging rod to it), for the coatings are only to be considered as conductors to the opposite surfaces of the jar; and the glass has only an attraction for a *certain quantity* of electricity, which, in its natural state, resides on each of its surfaces, and *no addition* can be made to the quantity naturally existing at *one surface*, but

* Figure 17.

by a correspondent *diminution* of the quantity naturally existing at the other. Hence, if a finger or other conductor be brought in contact with the outer coating of the insulated jar, or if a point be presented to it, its natural electricity will escape, and the jar will receive a charge.

Experiment 40. Take two Leyden jars of similar size, insulate one of them by placing it on a glass stand, and place the other on the table, with its knob at half an inch distant from the coating of the insulated jar, the knob of which should be placed at the same distance from the conductor of the machine; for every spark that passes from the *conductor* to the *knob* of the *first jar*, there will be a *similar spark* pass from the *coating* of the *first jar* to the *knob* of *the second;* and if they are successively discharged, the sound of the explosion and the brilliance of the light will indicate that they have each been charged to the *same degree.* Now, as the *second jar* was *charged* by sparks from the *coating* of the *first*, and as their *charges* were *equal*, it follows, that for every particle of electricity *added* to *one side* of coated glass, a corresponding particle *leaves* the *opposite surface.*

On this principle a jar may be charged by the transfer of its own natural electricity from one surface to the other, by insulating it and connecting its knob with the positive conductor and its outer coating with the rubber of the machine: electricity will be taken from the outer surface by the negative rubber, and conveyed to the inner surface by the positive conductor; so that the jar, though perfectly insulated, is charged by the unequal distribution of its natural electric fluid. This experiment is a satisfactory proof of the impermeability of glass to the electric fluid, for the conductor and the rubber of the machine are separated from conducting contact with each other only by the thickness of the glass jar, and a powerful accumulation of electricity takes place, which the contact of the thinnest film of conducting matter or the slightest fissure in the glass would have prevented.

When glass is charged, the attraction of its matter at the negative surface is necessarily exerted on the electric fluid accumulated at the positive surface, which is kept at a distance from it by the intervening thickness of glass; if the opposite surfaces are connected by a conductor, this distance is annihilated, the superfluous electricity rushes to the attracting surface and restores the equilibrium; such is the action of the discharging rod. But if no such connecting medium be applied, if the glass be thin, and its coatings separated by a sufficient extent of uncoated interval, the accumulation may be carried to such an extent as to break through the plate of glass to the attracting surface, producing a mechanical fracture, which, by rendering the glass permeable to electricity, prevents it from again receiving an electric charge. The thinnest glass, if sufficiently resisting, would produce the greatest accumulation, but a certain thickness is required to prevent the chance of fracture: Muscovy talc, in very thin lamina, is still highly resisting and susceptible of a considerable charge.

The different effects of the interposed substance on the quantity of electricity required to produce a given intensity, may be illustrated by coating equal surfaces of thick glass, thin glass, and a lamen of talc: if these be successively charged at a conductor with an electrometer attached, it will be found that the index of the electrometer will rise to any given degree with much fewer turns when glass is charged than with talc; and that the number of turns will be least of all with the thickest glass.— Hence it is evident that the common electrometers do not indicate the actual quantity of electricity, but merely its intensity, or tendency to an equilibrium by motion or explosion.

As the accumulation at the positive surface of charged glass depends on the attraction of the matter at the opposite surface, it follows that no discharge can take place unless these surfaces are brought in conducting communication; hence either side of a charged jar may be handled with impunity if it be first placed on an insulating stand; nor will the knobs of two differently electrified jars explode when brought together, unless their outer coatings are at the same time connected with each other.

Experiment 41. Take two equal jars, with a quadrant electrometer attached to the knob of each; place one of them in contact with the positive conductor of the machine and the other with the negative conductor;* when the ma-

* It is to be recollected, that when *one* conductor of the machine only is used, a chain or wire is to be suspended from

chine is turned both jars will charge, and to the same height, as may be seen by the receding index of each electrometer: remove the jars from the machine and place them on two separate insulating stands; connect their knobs by an insulated discharging rod; no explosion will ensue, although they are oppositely electrified; for their electricities depend on the attraction of their outer surfaces. which, in this insulated state, have no means of communication. Connect the outer surfaces by a wire or other conductor, and repeat the experiment; an explosion will take place, and both jars will be discharged.

Experiment 42. Place a jar, with an electrometer on its knob, in contact with the positive conductor, turn the machine until the index rises to 60; remove the jar and place it in contact with the negative conductor; on turning the machine the index will fall, and after a few turns the jar will be unelectrified, but if the turning be continued the electrometer will rise again with a negative charge; when, by remov-

the other to the ground; but when *both* are used together, the chain or wire is to be entirely removed.

ing it to the positive conductor, it may be again unelectrified.

The charge of any jar may be divided into equal and definite parts: by connecting its inner and outer coating with the inner and outer coatings of an unelectrified jar of the same size and thickness: the charge will be equally divided between them; and by repeating this process, the quarter, eighth, sixteenth, or any aliquot part of the original charge may be obtained.

Experiment 43. Hold a clean and dry pane of glass by one corner, and pass it before a ball connected with the positive conductor of the machine, so that the ball may successively come in contact with every part of the middle of the pane of glass, whilst the finger or any uninsulated substance is opposed to it on the opposite surface: by this process the glass will be charged. Apply the discharging rod to the opposite surfaces, an explosion will ensue; make the contact with the discharging rod again in another part of the surface, another explosion will be procured; and in this way many are sometimes obtained in succession. Repeat the experiment of charging, and then place the charged pane between two plates of metal of about half its size, on the application of the discharging rod but one explosion will be procured, but it will be louder and more brilliant than those procured from the uncoated pane. Hence it is seen that the use of the metallic coating is to *connect* the effects of every portion of the surface of the jar, so that it may be *charged* or *discharged* by the simple application of the machine or discharging rod to *one portion* of its surface.

Experiment 44. Place an uncoated jar beneath the conductor of the machine, and suspend a chain from the conductor so as to hang in the centre of the jar; on turning the machine the chain will move round, and apply itself in succession to every part of the internal surface of the jar, which by that means receives a charge. Apply the discharging rod, and the chain will return over the parts with which it has been in contact, and thus by a few of its revolutions the jar will be discharged.

Experiment 45. Take a Leyden jar, coated on the inside as usual, but with a coating of only one inch high on the outside; during the charge and discharge of this jar, ramifications of electric light will be seen on the outside, which are occasioned by the natural electric fluid of the outer surface passing from one part to another during its departure and return.

Experiment 46. Procure a jar with a double set of moveable tin coatings, either of which may be adapted to it at pleasure; the outer coating being a tin can large enough to admit the jar easily within it; and the inner coating a similar can sufficiently small to pass readily in the inside of the jar. The charging wire of the inner coating should be surrounded by a glass tube covered with sealing-wax, to serve as an *insulating* handle, by which the inner coating may be lifted from the jar when that is charged without communicating a shock to the operator. Arrange the jar with its coatings and charge it; it will act in every respect as an ordinary coated jar. Charge the jar, and without discharging it, remove the inner coating by its insulating handle; if this coating, when removed, be examined, it will be found not at all, or but slightly electrified: lift the jar carefully from within its outer coating, and examine that; it will evince no signs of electricity. Fit the jar up with the other pair of moveable coatings that have not been electrified, and apply the discharging rod; an explosion and spark will ensue; proving that the accumulation is retained by the attractive power of the *glass*, and that the *coatings* are consequently only useful as *conductors* to the charge.

The comparatively low intensity of any given quantity of electricity accumulated in the Leyden jar, when compared with a similar quantity disposed on an insulated conductor, occasions the jar to retain its electric state for a very long time under favourable circumstances: this fact was observed soon after the discovery of the instrument; and it has been since found, that if the surfaces are well separated from each other, the charge may be retained for many days or even weeks. The charge is usually dissipated by the *motion* of particles of dust, or other conducting substances in the atmosphere, from *one* of the *coatings* to the *other*, or by the uncoated interval becoming *moist*, and losing its *insulating* power; consequently a jar will retain its charge longer in dry than in damp weather: but the influence of external causes may be partly prevented by a particular construction. Coat the inside and outside of a narrow necked phial, and cement into it a glass tube long enough to reach to the bottom of the phial and extend an inch above its neck; the inside of this tube should be covered with tinfoil to rather more than half its length from the bottom, and its lining should be connected with the inner coating of the jar. A wire with blunt ends and half the length of the tube is to be placed loosely within it, and the top of the tube closed with a smooth brass cap. By this arrangement the contact of the atmosphere with any part of the interior coating is entirely prevented, and the conducting communication for the charge is procured by inverting the phial, which occasions the loose wire to form a temporary connection between the inner coating and the brass cap of the tube. When the charge has been communicated, the phial is to be set upright; the loose wire falls within the coated part of the tube and cuts off all *conducting* communication with the brass cap and the atmosphere. The phial will then retain its charge, and may be set by, or carried in the pocket until wanted; when being inverted, and a communication established between its outer coating and the brass cap, the discharge will be effected.

The simple operation of covering the uncoated part of the phial with melted sealing-wax, or with varnish, prevents the deposition of moisture upon it, and consequently tends also materially to prevent the dissipation of the charge.*

The properties of the Leyden jar, which have been developed by the preceding experiments, may be further illustrated by the very numerous varieties of them described in many experimental essays on the subject. The extent to which the modification of these experiments has been carried, may have arisen, in some instances, from an attempt to establish particular opinions, or to oppose others; and has been rendered necessary by the apparent anomalies which at first seemed to oppose any reference of the several phenomena of electricity to simple general principles. The supposed existence of a

* The surface of glass may be coated with sealing wax, by warming it gradually before a fire until it is hot enough to fuse the wax, a stick of which is then to be quickly rubbed over its surface. If varnish is used it may be applied with a flat camel-hair pencil, the glass being previously warmed. It is almost unnecessary so say, that any varnish that may be used for this or other electrical purposes, should be perfectly free from clamminess.

repulsive power, as a property of the electric fluid, has also tended materially to confuse our ideas on the subject; as, in order to explain the separation of *negatively* electrified bodies, it was necessary also to imagine that the particles of common matter were equally repulsive of each other; a supposition which is contrary to experience. Hence the fine superstructures of Mr. Cavendish,* and of Æpinus,† are considerably reduced in value by the hypothetical basis on which they are founded.

The mutual *attraction* of the electric fluid and common matter; the *elasticity* of the former and its tendency to the *surface* of the bodies with which it combines; the *different conducting* faculty of various substances; and the *alteration* of their *natural* attractive powers by contact, friction, expansion, or other *change* in their natural arrangements; are little more than simple expressions of the facts we observe: yet these simple principles have supplied an adequate solution of the phenomena we have yet considered, and are equally applicable to other apparent varieties of electrical action.

* Phil. Trans. vol. lxi. p. 584.

† Tentamen Theoriæ Electricitatis et Magnetismi.

In the Leyden jar it has been observed, that the *addition* of electricity to *one surface* is constantly attended by the *loss* of electricity from the *opposite surface*;* and this transfer, it has been shewn, is essential to the charge, which cannot take place without it. There are a variety of analogous phenomena, some of which it will be proper to consider in this place.

And first, with respect to the jar itself; it must follow, from the preceding principle, that during the process of charging, *both* surfaces of the jar evince the *same* electrical state: for, suppose the inner surface to be positive, it will have a tendency to give electricity to unelectrified bodies; and this is precisely what the

* That is, when, by the jar being *uninsulated*, a sufficiently extensive reservoir is provided for the reception of the *displaced electric fluid*; for, if the jar be *insulated*, a very small quantity only of electricity can be added to either surface, and that addition, by its action on the attractive power of the glass, occasions a similar portion to be released from its natural state of combination on the opposite surface: so that *both sides* of the glass evince signs of *positive electricity*. Such, therefore, is a case of *communicated* electricity; and, *it should be recollected*, that *such cases* are *distinctly separated* from the *charge*, and all analogous phenomena, in which the *natural quantity* of electric fluid is neither *increased* nor *diminished*, but is merely *unequally distributed*.

outside must do before it can become negative: but the inner surface appears positive, because the positive conductor is *adding* electricity to it faster than it can appropriate the attraction of the opposite surface to its increased quantity; and the outer surface appears so because its diminished attraction causes its natural electricity to leave it. Consequently, as soon as the turning of the machine is discontinued, the outer surface having *lost* a portion of its natural electricity, must be negative; and the inner surface, which has *increased* its original quantity, must be *positive;* although they appeared similarly electrified during the process. And the converse of this must be the case when the jar is charged negatively.

A plate of air, or any other nonconductor, may be charged in the same manner as a plate of glass; but as air is more readily displaced by electricity, in consequence of the mobility of its particles, a thicker stratum of it must be employed. The usual form of the experiment is to employ two circular disks of wood covered with tinfoil, and well rounded at the edges, having a diameter of from two to four feet. One of the boards is to be placed flat upon a table, and the other being suspended by a silk cord from the ceiling, is adjusted so as to hang parallel over its surface, and at the distance of an inch or an inch and a half from it. The upper insulated board being connected with an electrical machine, the stratum of air between the boards becomes charged, and will communicate a shock if the upper and lower one be touched at the same time with opposite hands. The shock produced in this way is considerably less violent than that from an equal surface of coated glass; for the distance of the coatings is of necessity much greater, and the medium between them less perfectly insulating: and this last circumstance operates so rapidly when the charge is high, that its maximum of effect cannot be obtained but by making the discharge whilst the machine is still in action. If the discharge be not made, spontaneous explosions from one disk to the other, through the intervening plate of air, will occur at intervals, as long as the electrization of the upper disk is continued.

Analogous to the process of the last experiment is the production of the electric spark under ordinary circumstances: when any conducting substance, in its natural state, is presented to a positive conductor, its matter is attracted by the proximate electricity of the positive body, and the electric fluid before *diffused* over its surface *retires* to the most *remote* parts: if the presented substance be *insulated*, it will therefore become *negative* at the surface *near* the positive conductor, and *positive* at the surface which is most *remote;* and if it be sufficiently light or pendulous, it will move toward the positive conductor until a spark occurs between them. If, instead of an *insulated* body, any conducting substance in *connection* with the *ground* be presented to the positive conductor, its presented surface will become more highly negative, since its natural electricity has *unlimitted* room to recede: hence *uninsulated* bodies are attracted at greater distances, and receive stronger sparks than the largest of such as are insulated.

It is thus seen, that whenever sparks or attractions are produced, the bodies between which they pass are necessarily in opposite states of electricity, and are therefore analogous to the coatings of a Leyden jar, and serve indeed as coatings to the plate of air by which they are separated: the force of the spark will therefore be influenced by the extent of the insulated conductor, and the perfect connection with the ground of that which is opposed to it When a very large conductor is attached to an electrical machine, the spark from it may be made equivalent to a shock, by any individual standing on a wire connected with a well or water pipe, and receiving the sparks from the conductor on a large brass ball held in the hand.

From this tendency of electrified bodies to produce an unequal distribution of the natural electricity of all such substances as are brought sufficiently near them, some curious phenomena result; and the action of some of the most interesting instruments are dependent entirely on their operation. Of this kind are the Electrophorus, and the Condenser; two very remarkable sources of electrical accumulation, invented by professor Volta.

The electrophorus consists of two circular plates of metal, or of wood covered with tinfoil and well rounded at the edge; these are called the conductors: between them is placed a resinous plate, formed by melting together equal parts of shell-lac, resin, and venice turpentine, and pouring this mixture, whilst fluid, within a tin hoop of the required size, placed on a marble table, from which the plate may be readily separated when cold. This resinous plate should be about half an inch thick: it is sometimes made by pouring the fluid mixture on one of the conductors, which is then formed with a rim for that purpose. The conductor on which the resinous plate is placed is called the lower conductor, or sole; and that which is placed upon the resinous plate the upper conductor, or cover: this last is always furnished with a glass, or other insulating handle; and when the electric state of the lower conductor is to be examined, the whole apparatus is placed on an insulating stand.—See Figure 18.

Experiment 47. Rub the upper surface of the resinous plate with a piece of dry fur (cat's skin is the best); it will be excited negatively. Place the upper conductor upon it, and then raise this last by its insulating handle; it will be found to exhibit very faint, if any, electrical signs. Replace the conductor, and, whilst it lies on the surface of the excited plate, touch it with a finger or other uninsulated conductor, and then raise it again by its insulating handle;

it will now appear positively electrified and afford a spark: if it be then replaced on the resinous plate, touched and again raised, another spark will be procured; and this process may be repeated for a considerable time without any perceptible diminution of effect.

The uniform result of this alternate contact and separation of the conductor, without any new excitation of the resinous plate, evinces that the actual electric state of the latter is not destroyed by that process; and the necessity for the connection of the conductor with the ground before it is raised, proves that the acquired electricity is derived from that contact.

The nonconducting faculty of the resinous surface, and the imperfect contact the flat conductor forms with it, precludes the transmission of electricity of low intensity from one to the other; when in contact, they are therefore only to be considered as *very near* each other: now it has been seen, that when an *insulated conductor* is brought *near* an electrified body, the natural distribution of its electric fluid is disturbed, and the conductor becomes *oppositely* electrified at the *anterior* surface, and *similarly* electrified at the *posterior* or *remote* surface; consequently the *cover* of the electrophorus, when laid *upon* the resinous plate, which is *negative*, will have its natural electric fluid determined *toward* that plate, and it must then appear also *negative*, and will *receive* electric fluid from any conductor brought near it: this *increased* capacity, arising only from its extreme *proximity* to the resinous plate, *ceases* when the cover is *raised* by its insulating handle; and the additional electricity it has received is then given off in the form of a spark to the first conducting substance it approaches.

This explanation is by no means what would be inferred from a superficial view of the phenomena, for the spark that passes *to* the cover whilst it is on the resinous plate is less considerable in its appearance than that which passes *from* the cover when it is raised, although it is here stated that they consist of the *same quantity* of electricity: it is therefore necessary to shew that such is really the case.

Experiment 48. Place the cover on the excited plate by means of its insulating handle, and bring the knob of an unelectrified Leyden jar in contact with it, then touch the cap of an electrometer with the knob of the jar, and it will diverge with negative electricity. Raise the cover, and present the knob of the jar to it, a strong spark will pass; bring the knob of the jar in contact with the negatively electrified electrometer, its divergence will be *exactly* destroyed: now this effect could only be produced by an *equal* quantity of positive electricity; for had it been *more*, the electrometer would have separated again with the *excess*, and have remained slightly positive; and had it been *less*, the original divergence could not have been *wholly* destroyed, and the electrometer must in consequence have remained slightly negative. When the lower conductor of the electrophorus is also insulated, it evinces electrical signs: if the excited plate only be placed on it, it is negative; but when the cover is placed on the plate, its state changes to positive; and when the cover is raised again, returns to negative: so that the opposite coatings of the plate are always in opposite states of electricity, which might be expected, since their arrangement is similar to that of the coatings of a Leyden phial, from which instrument the electrophorus differs only by combining the power of an electrical machine with the properties of the jar; the excited surface of the resinous plate being a permanent source of variable attraction in its contiguous conductors, and the approximation or removal of one of these modifying the influential power of the excited surface on the opposite one.

The power of a well-constructed electrophorus is sufficient to adapt it as an occasional substitute for the electrical machine: since about 20 sparks from the raised cover of the instrument, given in succession to the knob of a moderate sized Leyden phial, communicate rather a strong charge to it; and the permanent action of the excited plate admits the frequent repetition of similar experiments.

The Condenser, is an instrument in which Professor Volta has applied the principle of approximated surfaces, to the detection of such slight electrical changes as are not appreciable by the most delicate electrometers. The preceding details have shewn, that any insulated conductor, opposed to one that is not insulated, has its capacity of electrical change increased by that proximity, and is more susceptible of an increased or diminished quantity of electric fluid than when freely insulated; because in the state of approximation a much more considerable *change* will be required to produce the *same intensity*, or tendency to equilibrium.—Now, was the contiguity of the opposed plates permanent, no advantage would be obtained; for the principle which renders the insulated plate susceptible of more extensive electrical change, also prevents it from rendering that change evident: it is therefore essential that the plates should be so arranged as to admit of alternate proximity and separation.

The most simple condenser may be formed by placing three small spots of sealing-wax, at equal distances, on the lower face of the cover of an electrophorus, to serve as short insulating legs by which it may be supported at the distance of about a twelfth of an inch from the surface of a smooth and even table. If a Leyden jar be now charged, and afterwards discharged, so as not to affect an electrometer, and its knob be then placed in contact with the condenser resting upon the table for a few seconds, the very small residuum of electricity remaining in the jar will be absorbed by the condensing plate; and when this is raised from the table it will affect the electrometer with the same electricity as that with which the jar was charged.

The most improved condensers have the insulated plate fixed on a glass pillar, and the uninsulated plate supported by a brass wire with a joint and stop: the plates are parallel to each other, and when electricity is to be communicated to them they are situated at the distance of the thickness of a card from each other; the uninsulated plate is then drawn back, and the intensity of the insulated plate displayed.—See Figure 19.

The power of the condenser, thus constructed, is not always sufficient to manifest very slight effects: Mr. Cavallo extended its application, by transferring the electricity of the first condenser to the insulated plate of another of smaller size; this small plate is now usually attached to the cap of a gold leaf electrometer, and a similar plate is opposed to it by a jointed wire connected with the foot of the instrument. See Figure 20.

The use of these combined condensers affords a means of detecting very slight electrical changes; and so obvious is the importance of this property, that many eminent electricians have bestowed considerable labour in attempts to produce more perfect and delicate arrangements for its application to purposes of research. Of this kind are the doublers of electricity, invented by Mr. Bennet and Mr. Nicholson;* the multiplier of electricity, contrived by Mr. Cavallo; † the electrical spinning instrument of Mr. Nicholson; ‡ and a double multiplier of Mr. Wilson's: § contrivances of considerable ingenuity, by which the powers of the simple condensers are far exceeded: but, unfortunately, the increased sensibility of these instruments is attended by a tendency to produce the electrical states spontaneously, and the equivocal results they consequently afford, is a very considerable abridgment of their utility.

The various phenomena that have been now considered, include the most important diversities of electrical action; they cannot be contemplated without perceiving a distinction between the *causes* of the electrical appearance of different insulated conductors; for they display

* Phil. Trans. vol. lxxvii. p. 288. and vol. lxxviii. p. 1. 403.

† Cavallo's Complete Treatise on Electricity, vol. iii. p. 99.

‡ Nicholson's Journal, 4to. vol. i. p. 16.

§ Nicholson's Journal, 8vo. vol. ix. p. 19.

two *separate* methods of exciting those appearances. First, by an actual *alteration* of the *natural quantity* of the electric fluid the conductors contain; and, secondly, by its *unequal* distribution in their *proximate* and *remote* parts. The first method can only be employed by *conveying* electricity *to* or *from* the conductor; it is consequently called *communicated* electricity, or the electricity of contact, and remains *permanent* so long as the insulation is maintained. The second method obtains whenever an insulated conductor is brought *near* an electrified body; the *presented* surface obtaining a *contrary* electrical state, and the *remote* extremity being *similarly* electrified, whilst a neutral unelectrified point exists between them: but these electrical states being the *mere* effect of a *disturbed electrical arrangement*, are *only* permanent *whilst* the *proximity* of the electrified body to the insulated conductor is *continued*, provided its insulation has been perfect. Such phenomena are classed under the general term electrical influence; and the positive and negative states so produced are called the electricities of position, or approximation, and by some writers induced electricity.

PART II.

OF THE MECHANICAL AND CHEMICAL AGENCIES OF ELECTRICITY.

CHAP. I.

Instruments required for the Application of the Electric Power to the Purpose of Experiment.

THE apparatus, hitherto described, is adequate to the production and accumulation of electrical effects; but when the influence of the electric fluid on the bodies through which it is made to pass is to be investigated, some contrivances are required for its accurate and convenient application to that purpose.

The form in which accumulated electricity is most usually employed, is that of a charge; hence a variety of Leyden jars are required; for, although the same *intensity* of charge may be obtained with every jar of equal thickness, the *quantity* of electricity will be in proportion to the *extent* of surface; and the quantity is a consideration of importance when good conductors are employed to transmit it.

Very large jars cannot be obtained; the largest I have yet seen is one in my possession, which is eighteen inches diameter and two feet high: the coating on the outside of this jar exposes a surface of about six square feet, which is by no means sufficient for all purposes: when great electric power is required, it is therefore usual to combine several jars together, so that they may be charged or discharged at once as a single jar. Such a combination is called an Electrical Battery; and it is obvious that, by increasing the number of jars, any required extent of coated surface may be obtained.

The structure of an electrical battery should be simple, for its parts are occasionally deranged during its use; one or two jars sometimes break by a spontaneous explosion, and until they are removed and replaced by others, the battery will be useless. The jars are usually placed in a box with thin partitions, to prevent their mutual contact. The bottom of the box inside is covered with a trellis of wire, or with tinfoil, on which the coated bottoms of the jars rest;

and their outer coatings are consequently in conducting communication with each other. If there are twelve jars, they may be placed in three rows of four each; every jar having its charging wire terminated by a smooth ring instead of a ball. A brass rod, with balls at its extremities, being passed through the rings in each row will connect the inner coatings of four jars; and the rods of the three rows may be connected together by laying two shorter rods from one to the other: as the short rods are moveable, either four jars, eight jars, or the whole battery may be employed at pleasure.*

A battery is charged and discharged in the same manner as a single jar, namely, by bringing the charging wires of its inner coating in contact with the positive conductor of the machine whilst the outer coating is in conducting communication with the table, and, after the charge has been communicated, connecting the outer and inner coatings by the discharging rod or any other conductor. The communication from the machine to the battery may be made by jointed brass rods, or by a thick copper wire, care being taken that no points or

* Figure 21.

edges are exposed, which would tend to dissipate or weaken the charge.

When the uncoated interval of a Leyden jar is very clean and dry, no very considerable charge can be given to it before an explosion takes place from coating to coating over the dry glass; and, as the charge is lost by this occurrence, it becomes a source of inconvenience, and severely so when large batteries, which it requires a long time to charge, are employed. This tendency to spontaneous explosion may be much diminished by covering part of the uncoated interval with any imperfect conductor; such effects have been produced by slightly soiling the glass with handling it when the hand is in a state of perspiration, by breathing slightly on one of its surfaces, by placing a wet sponge within the jar, or by slightly oiling its surface: but these methods are not permanent in their effect, and have been consequently superseded by a simple arrangement more recently proposed; which consists in pasting a slip of writing-paper, of an inch broad, on the inner surface of the jar, so as to cover the uncoated interval to the height of half an inch above the upper edge of the inner coating. The action of this, and of the other means that have been employed for the same purpose, appears to consist in a gradual diminution of the intensity of the charge at that part from which it has the greatest tendency to explode, by an extension of the charged surface through the medium of an imperfect conductor.

The height of the uncoated rim of the jar should be proportioned to the charge it is intended to resist; with small jars, 2 inches, or $2\frac{1}{2}$ inches is sufficient, the coatings being then separated by an interval of 5 inches: with larger jars a rim of 3 inches will be usually adequate, if they are fitted up with an interior paper band.

The uncoated part of the jars in a battery are sometimes varnished, which prevents the deposition of moisture, and is of advantage if the varnish be good: when varnish is employed, the paper band is indispensable, for the tendency to spontaneous explosion is much increased by the uniform dry surface the varnish presents. If the jars are not varnished, the exterior of the uncoated rim must be kept dry and free from dust.

The jars in a battery should not be very thin, for the chance of fracture is greater when a battery is employed than with a single jar, in proportion to the number it may contain. If the jars are but moderately thick, it will be of advantage to interpose a thickness of writing-paper between the coating and the glass, which may be easily effected by pasting the tinfoil first on paper and afterwards applying this combined coating to the glass. The metallic coatings are thus placed at a greater distance from each other, and the chance of fracture is diminished.

Next to the sources of electrical accumulation, it becomes necessary to consider the means of estimating and directing the power we employ; since the same jar or battery is susceptible of various degrees of charge. The application of the quadrant electrometer to this admeasurement has already been described; it is placed on the conductor, and consequently in contact with the internal coating of the jar or battery, and indicates, by the rise of its index, the intensity of the charge conveyed.

Lane's discharging electrometer is somewhat different in principle: it consists of two balls of equal size, one connected with the inside of the jar, the other insulated opposite to the first, but capable of being placed either in contact with or at any distance from it. The insulated ball is connected with the outer coating by a wire: it is therefore a vehicle for the discharge, which will take place sooner or later in proportion to the distance at which the balls are placed. The principal imperfection of this electrometer arises from the occasional intrusion of particles of dust or other light conducting matter between the balls, by which the indications of the instrument are rendered fallacious.—Fig. 22 represents a jar fitted up with Lane's electrometer.

The most useful electrometer for jars and batteries is that constructed by Mr. Cuthbertson: it consists of a metal rod, about 13 inches long, terminated by balls, and balanced on a knife-edged centre in the manner of a scale-beam. One arm of the balanced rod is graduated, and has a slider upon it, which, when placed at different distances from its fulcrum, loads the arm with a proportionate weight from one grain to 60. The graduated extremity of the balance rests upon a similar brass ball, which is supported by a bent metal tube from the same insulating stand; and at four inches below the opposite extremity another insulated ball is placed, which is to be connected with the outside of a jar or battery. Now, if the metallic support of the balance be connected with the conductor, or the inner coating of the jar, and this last be electrified, there will be an attraction between the extremity of the balance and the lower insulated ball, because they are connected respectively with the opposite surfaces of the jar; and when the force of this attraction exceeds the weight with which the opposite arm is loaded, the attracted arm of the balance will descend, and discharge its electricity on the lower insulated ball. The power of the attraction is always proportioned to the intensity of the charge; and as, in this instrument, the attraction has to overcome a resistance proportioned to the weight with which the balance is loaded, that weight becomes a proper comparative measure of the intensity of any required charge.—Fig. 23 represents this instrument surmounted by the quadrant electrometer, which is useful to indicate the *progress* of the charge, as that is not shewn by the action of the balance electrometer itself.

All these instruments indicate only the *intensity* of the accumulated electricity, or its deviation from a state of natural distribution: the *quantity* can only be inferred from the comparative *extent* of the charged surface, or estimated by an examination of its effects, and is therefore by no means accurately appreciable.

The discharge of a charged jar is (as it has been already stated) effected by connecting the inner and outer coatings with some conductor: this process is called forming a Circuit; and any substance interposed between two parts of the connecting conductor, or between it and one of the coatings of the charged surface, is said to be introduced into, or placed in, the circuit. The most simple way of effecting this is to place the body, through which the charge is intended to pass, in contact with the outer coating, and putting one knob of the discharging rod upon it, bring the other quickly toward the charging wire of the jar; the electric fluid is thus constrained to pass through the interposed body, or over its surface. The rapid approximation of the discharging rod is necessary to ensure the full force of the charge, a part of which would otherwise pass silently without producing any perceptible effect.

The formation of a circuit has been defined (when treating of the action of the discharging rod) to be the annihilation of all distance between the metallic coatings; it will consequently be produced most rapidly through the best conductors, and prefer the shortest possible course; hence if a person hold a wire between his hands, whilst with one he touches the coating of a charged jar and brings the other to its knob, he will feel no sensation but at the points of contact; for the electric fluid having the choice of two circuits prefers the best conducting, and passes through the wire without affecting the body of the person that holds it. If he, however, substitute a piece of wood for the wire, a shock will be felt; for dry wood is a worse conductor than the animal fluids, and the charge having two circuits, passes through that which affords it the easiest passage.

This fact is well illustrated in every method of communicating the electric shock, which is only felt along the muscles in the most direct line that enters into the circuit. Let A, B, C,

D, E, and F, hold each other by the hand, A having the outer coating of a charged phial in his hand, the knob of which is to be touched by F; each individual will be shocked in the same manner, and at the same time, the sensation reaching from hand to hand through the arms, and, if the charge be strong, across the breast. These are the parts of the body that enter directly into the circuit. Vary the experiment; A holding the phial as before, and touching the right foot of B with his left foot. The left foot of B to touch the right of C, and so on to F, who is to complete the circuit by touching the knob of the phial with his left hand: A will be shocked in his right arm and left leg, B, C, D, and E in both legs, and F in the right leg and left arm; so that the charge passes in the most direct line from one point of contact to another.

The extent to which a charge may be conveyed by good conductors is remarkable. At a very early period the Abbé Nollet communicated an electric shock from a small phial to 180 of the king's guards, and afterwards to a convent of Carthusians; and the sensation was felt by all the persons forming that extensive circuit at the same moment.

Experiments have been made to ascertain the velocity with which an electric charge moves, but hitherto without success. Dr. Watson and some other members of the Royal Society formed a circuit, with iron wires, of upwards of four miles extent, but the charge required no appreciable time in passing through this lengthened interval.*

The tendency of the charge to pass through the best conductors, offers a measure of conducting power; for if various substances of the same length and size are introduced at once into a circuit, that through which the electric fluid passes is the best conductor. Or if they are introduced successively, that which conveys the

* There is some doubt as to the accuracy of these experiments; they were made at a very early period, and have not, I believe, been repeated since the improved state of the science has afforded the means of effecting such experiments with precision. Metals, although the most perfect conductors we have, oppose some resistance to the motion of electricity, and a charge will even prefer a short passage through air to a circuit of 20 or 30 feet through thin wire. It is therefore rather uncertain that the charge of a small phial has ever passed through an interval of four miles.

charge most completely may be considered the best conductor.

To transmit the charge with more certainty and precision, an ingenious apparatus was contrived by Mr. Henly: it consists of a mahogany board, 14 inches long and 4 wide, having a socket fixed in its centre, to which may be alternately adapted a small table with an ivory top, or a mahogany press. Two wires, sliding in spring tubes, and mounted on universal joints, are fixed to the top of two glass pillars, which are cemented near the extremities of the mahogany base at equal distances from the central socket. The body through which the charge is intended to be passed is placed on the table, or screwed in the press, which is then adjusted in its socket. The sliding wires, which are moveable in every direction, are then brought in contact with its opposite sides, and one of them being connected with the outside of a jar or battery, and the other with the discharging rod or a discharging electrometer, the charge is determined through it with great accuracy.—The instrument is represented by Fig. 24: it is called the Universal Discharger.

When electrical sparks are intended to be passed through various substances, their action may be rendered uniform by receiving them on an insulated ball, in contact with which the subject of experiment is to be placed, and its opposite extremity connected with the ground; the insulated ball should be of the same height with the conductor of the machine, and, being fixed on a separate stand, may be placed at any required distance from it; by which means the sparks may be made stronger or weaker at pleasure.

When the electrical spark is to be passed through different fluid or elastic mediums, these substances are enclosed in glass tubes, and two wires are inserted through the opposite sides or ends of each tube, so as nearly to meet in its centre; between these wires the spark is to be passed, and an experiment of this kind may be continued for any time without opening the vessel.—See Figures 25 and 26.

When fluids are acted upon in this way, they are sometimes placed in a tube, closed at one end, through which a platina wire passes, and is continued through the centre of the tube until it comes within a short distance of its open extremity: the tube being inverted in a brass dish, sparks may be transmitted from the point of the wire to the bottom of the brass cistern. See Fig. 27.

It is evident that apparatus of this kind may be modified to any extent, and that the construction of it is, for the most part, exceedingly simple: the agencies of electricity, in altering the forms or characters of other matter, are only exerted when it passes from one body to another; so that an interrupted circuit is essential to every electro-mechanical and electro-chemical apparatus.

Tubes of glass, wires of different metals, corks, and a few other materials, are adequate to the construction of an endless variety of electrical machinery, and the proper direction of such resources is constantly followed by useful discovery. Mechanical dexterity is therefore essential to the character of an electrician, since his progress will be in proportion to the facility with which he can adapt the objects around him to new inquiries. He cannot deviate from the beaten track of his predecessors without the aid of new combinations; and when the supply of these is derived from his own industry and ingenuity, the ardour of his pursuit will be unimpeded by the delays or mistakes of others; and the projection of any required improvement may consequently be followed by its immediate consummation.

CHAP. II.

Mechanical Effects of Electricity.

THE transmission of the electric fluid from one body to another is always attended by some mechanical effect; when its motion is slow, light substances are moved by it or currents of air produced; when its motion is rapid, light is evolved and a sharp sound ensues. The sound is produced by the sudden collapse of the air, which has been displaced by the passage of the electric fluid; and it is consequently greater in proportion to the quantity and intensity of the charge. Hence when different sized jars are charged to the same degree, and then successively discharged, the explosions produced will be louder in proportion as the jars are larger; and the effect afforded by a battery of extensive surface will be that of a comparatively violent report.

The immediate consequence of the passage of an electric charge through any substance, appears to be an expansion or removal of the particles directly in its course, and a consequent compression of those by which they are surrounded: so that the result of an electrical explosion is usually some evidence of the action of an expansive power.

Experiment 49. Place a card, or the cover of a book, flat against the outer coating of a Leyden jar, exposing about a square foot of coated surface; put one extremity of a discharging rod against the card, and bring the other extremity to the knob of the jar: the charge will pass through the card and perforate it, producing a small bur or protrusion on the side next the discharging rod, and a larger bur on the side which was in contact with the coating of the jar. By employing a battery, a quire of strong paper may be perforated in the same manner; and such is the velocity with which the electric fluid moves, that if the paper be freely suspended, not the least motion is communicated to it.

Experiment 50. Put a piece of dry writing-paper on the table of the universal discharger, and, removing the balls from the ends of the sliding wires, press them upon the paper at the distance of about two inches from each other;

pass a strong charge from one wire to the other, and the paper will be torn in pieces. If a number of wafers are placed on the table instead of the paper, they will be dispersed in a curious manner, and many of them broken to pieces when the charge is passed through them.

Experiment 51. Drill two holes in the opposite ends of a piece of wood which is half an inch long and a quarter of an inch thick; insert two wires in the holes, so that their ends within the wood may be rather less than a quarter of an inch distant from each other: pass a strong charge through the wires, and the wood will be split with violence. Loaf sugar, stones, and many other brittle nonconductors, may be broken in the same way, if a sufficiently powerful charge be employed.

Experiment 52. Introduce two wires into a soft piece of pipe-clay, and pass a strong shock through them: the clay will be curiously expanded in the interval between the wires. The experiment will not be successful if the clay be either too moist or too dry.

Experiment 53. Insert two wires through corks in the opposite ends of a small glass tube; let the distance of the ends of the wires be about half an inch: fill the tube with water, and pass a moderate charge through it; the tube will be broken and the water dispersed.

The expansion of fluids by electricity is indeed very remarkable, and productive of some singular results. When the charge is strong, no glass vessel can resist the sudden impulse. Beccaria inserted a drop of water between two wires in the centre of a solid glass ball of two inches diameter; on passing a shock through the drop of water the ball was dispersed with great violence. Mr. Morgan succeeded, by the same means, in breaking green glass bottles filled with water, when the distance between the wires that conveyed the spark and the sides of the glass exceeded two inches. With but a moderate charge I have, in this way, broken glass tubes the thickness of half an inch in the sides, and with a bore of the same diameter.

Experiment 54. Place a piece of plate glass, about an inch square and half an inch thick, within the press of the universal discharger, or lay it flat upon the small table and press it by a weight; set the points of the sliding wires opposite to each other, and against the under edge of the glass, so that the spark may pass beneath it: the charge of a large jar, transmitted in this way, rarely fails to break the glass.

Experiment 55. Form a small mortar of ivory, with a cavity of half an inch diameter and an inch deep; insert two wires through the sides of the mortar, so that their points within its cavity may be separated by an interval of a fourth of an inch; fit a cork cap so as to close the aperture as accurately as may be without friction: when a strong charge is passed through the wires, the air withinside the mortar is suddenly expanded, and the cork is projected to a distance with some violence.

Experiment 56. Let a spherical cavity be turned in a piece of ivory capable of receiving the half of a light wooden ball; a small conical cell is to be made at the bottom of the spherical cavity, and two wires inserted through the sides of the mortar into it: if a drop of water, oil, alcohol, or ether, be put between the wires, and the ball placed over them in its cavity, a charge sent through the drop of fluid will convert part of it into vapour, and expel the ball with considerable force.

If a discharge be passed over ice, the surface is sometimes marked with spots, as if a hot chain had been laid upon it; if it be passed over snow, it divides the portion over which it passes; when it is taken over the surface of soft dough, a permanent depression is made in the track of the discharge; and when it passes through a green leaf, the leaf is torn to pieces. Hence it appears that an expansive effect is produced at every interruption of the metallic circuit, or when the electric fluid elicits light during its passage; and this expansive power produces a mechanical effect proportioned to the nature and resistance of the medium in which it occurs. Even the best conductors, when in a sufficient state of tenuity, are considerably expanded by electricity. Let a capillary tube of glass be filled with mercury, and pass a charge through it; the mercury will be expanded with sufficient force to splinter the glass tube.

Electric light is evolved at every interruption of a metallic circuit, even in conducting fluids when the charge of a jar is passed through them. This experiment may be made with water in a thick glass tube, having two wires within it with their ends almost in contact. A moderate charge will produce a bright spark; but no more power should be employed than is absolutely necessary for the purpose, as there is some danger of breaking the tube. A very low charge may be first employed, and if it is found insufficient, it may be gradually increased until the required power is obtained.

In oil, alcohol, or ether, the spark is more readily procured, as they are better insulators; but the expansibility of these fluids renders the experiment even more dangerous with them than with water. As the difficulty of eliciting electric light in any medium increases with its conducting power, a much higher charge is required to procure a spark in hot water than in cold: in saline fluids the difficulty is further increased; and in concentrated acids, light can only be procured when their volume is comparatively trifling.

The effect of some of these fluids on the striking distance through air is very remarkable. Dr. Priestley first remarked, that the explosion from a large battery would pass to a greater distance over the surface of water than in the free atmosphere. This fact is best exemplified in the following manner.

Experiment 57. Draw a line with a pen dipped in water on the surface of a strip of glass;

place one extremity of the line in contact with the coating of a Leyden jar, and at six inches distance upon the line place one knob of the discharging rod; when the jar is fully charged bring the other knob of the discharger to the ball of the jar, and the discharge will take place luminously over the six inches of water.

Experiment 58. With a pen dipped in sulphuric acid trace a line on a strip of glass, as in the former experiment, and place one extremity of it in contact with the outside of the jar; the ball of the discharger may be placed on the strip at twelve inches distance, and the electric fluid will pass as brilliantly over that interval as over the six inches of water.

In either of these experiments, if the strip of fluid be wider in any particular part, the light of the discharge will appear less brilliant in passing that portion; which may arise from the greater division of the fluid when passing over an extended conductor than over one that is narrow.

The great mechanical power evinced by the electric fluid in these experiments, is a strong proof of its claim to the character of a material substance; and the impermeability of nonconductors to it, when in their perfect state, is shewn by the mechanical injury they sustain when, by reducing them to sufficiently thin lamina, a strong electric charge is conveyed through them.

The hardest and most compact bodies may, by such means, be broken or perforated; but from the influence of surface, or other causes, their resistance is not always proportioned to their insulating powers. The perforation of solid nonconductors is effected with the least power when the discharge is confined to a single point, and prevented from dispersion by being surrounded by some nonconducting matter; as may be thus exemplified.

Experiment 59. Fill a small phial with olive oil, and introduce within it a pointed wire bent at right angles, so that by sliding through a cork placed in the neck of the phial the point of the wire may be made to rest against any part of its inside beneath the oil: suspend the phial by its wire to the conductor of an electrical machine, and place the knuckle or a brass ball near the outside of the phial and opposite to the point of the wire that is within it: a spark will pass from the point to the knuckle, and make a small hole in the glass. By turning the point round, or raising it higher or lower, many such holes may be made.

The point serves as an internal coating to a very small portion of the glass, and the charge being prevented from extending by the surrounding oil, the whole power of the machine is concentrated to that point, and consequently soon overcomes its resistance. Similar effects will always ensue when a large quantity of electricity is suddenly transferred to a comparatively limited surface.

Experiment 60. Charge a very large jar; connect its outside with one that is ten or twelve times smaller: make a communication between their inner coatings with the discharging rod, and the small jar will be broken; the quantity of electricity transferred to it, being beyond the proportion of its size.

To ascertain the resisting faculty of various substances, a pointed wire should be procured, and surrounded by a cylinder of wax or pitch, which, being softened, may be applied to the surface of a lamen of the substance to be tried, and will confine the action of the point to one part of that surface; the opposite side of the resisting plate is to be connected with the outside of a charged jar, and the discharge made through the point. In this way Mr. Morgan ascertained that of bees'-wax or sulphur, a plate of 3-10ths of an inch thick could be perforated by a high charge; whilst of plate glass 3-20ths of an inch was the greatest interval that could be overcome by his apparatus; and of shell-lac only 2-20ths of an inch were struck through. Hence, for its practical application, shell-lac appears one of the best nonconductors we have.

Many substances, that are tolerable conductors of electricity, may be also perforated by an electric charge, when its action is confined to one part of their surface; such is usually the case with the tinfoil coating of large jars, if, when they are highly charged, the discharge is made by touching the tinfoil only in one place with the discharging rod; the little spark that occurs at the point of contact fuses the tinfoil, and a slight adherence is almost always observed in consequence, between the knob of the discharger and the coating after every discharge. This property may be thus exhibited.

Experiment 61. Charge a large jar, and place a shilling or other piece of coin between the knob of the discharger and the coating of the jar: when the discharge is made the coin will be slightly soldered to the tinfoil by its fusion at the point of contact, and will remain adhering to the coating after the discharger is removed.

The mechanical effects of electricity have been employed to indicate the course of the electric fluid in the discharge, and thus to confirm the proposition that assumes positive electricity to be an accumulation of electric fluid, and negative electricity to be a deficiency; in opposition to the hypothesis first proposed by Du Faye, that positive and negative are two distinct electric powers.

It has been already shewn, that the phenomena yet considered indicate the agency of only one fluid; and the effects attendant on the transmission of the charge will be found to afford the same indication, although some of them have been differently interpreted by some inexperienced electricians. There is, probably, no science in which manual dexterity is more essential to successful inquiry than that of electricity; the facts that constitute it are also numerous and of diversified character, yet so dependent on each other, that until their mutual relations are understood, and all the intricacies of electrical action clearly comprehended, no just application of its principles can be effected. It is therefore, perhaps, not very surprising that the tyro, when he attempts to reason from a limited range of observation, is frequently led to adopt erroneous conclusions, which have no other foundation than in the inaccuracy of the experiments he may have made, and in his want of skill in the arrangement, classification, and comparison of phenomena.

Experiment 62. The direction of the electric fluid is rendered visible when a Leyden jar, which has been rendered slightly damp by breathing on it, is placed with its knob in contact with the positive conductor of the machine in a darkened room: when the jar is fully charged, if the turning of the machine be continued, the electric fluid will be seen to pass from the inner to the outer coating over the uncoated interval in luminous streams, producing an effect similar to that of water overflowing from the top of a vessel that is kept constantly supplied. If the jar be removed, and its knob placed against the negative conductor, the stream, when the jar is overcharged, will evidently pass in the contrary direction, that is, from the outer to the inner coating. A certain degree of dampness on the uncoated part of the glass is necessary, in this experiment, to prevent the discharge of the jar by spontaneous explosion, in which case the fluid passes too rapidly from one surface to the other to admit the ascertainment of its direction. If the moisture be not sufficient, divergent brushes of light pass from the positive surface at intervals, instead of the continuous streams before described.

Experiment 63. Let a small jar be charged positively on the inside; place it under the receiver of an air-pump: on exhausting the air, brushes of light will pass from the knob of the jar to its coating. Repeat the experiment with a jar charged negatively; the direction of the flashes of light will be reversed.

Experiment 64. Place a lighted taper between the wires of the universal discharger, they being at four inches apart, and the flame midway between them; connect the coating of a small charged jar with one wire, and bring its knob in contact with the other: if the charge be just sufficient to pass the interval without explosion, the flame of the taper will be constantly blown from the positive wire to that which is negative.

Experiment 65. Lay two very straight sticks of sealing-wax on the table of the discharger parallel to each other, so that the juncture of their rounded edges may form a groove; on this a large pith-ball is to be placed, and the wires of the discharger are to be arranged with their points in the direction of the groove, and at four inches from each other, the ball being equally distant from each. On passing a small charge from one wire to the other, the ball will be driven from the positive to the negative; and this effect will be constant if the terminations of the wires are pointed, which they should be for these experiments of transmission. If blunted wires are employed, the ball frequently vibrates between them, and apparently renders the result equivocal; but it should be recollected, that by employing knobbed wires, the *transmission* of the charge is *prevented;* and, as the wires are connected with the opposite sides of the jar, they must necessarily attract the ball alternately, as any other oppositely electrified conductors would do. Again, it may also be observed, that even with pointed wires the motion of the ball is not always in the supposed direction of the fluid; for if it be placed in *contact* with *either wire*, it will move *from* that wire as soon as the circuit is completed, whether the wire be in contact with the *positive* or *negative* side of the jar; but this, when attentively considered, proves nothing relative to the course of the fluid; for the ball becomes *electrical* by its *contact* with the wire, and consequently recedes from it toward the opposite surface, by which it is *then attracted*. That attraction is the cause of this apparent anomaly may be proved by making the experiment with a jar exposing about a square foot of coated surface, which is to be moderately charged, first with positive electricity and then with negative; its outside being connected with the wire toward which the ball is to move, and the circuit completed with the discharging rod, by connecting the opposite wire (against which the pith-ball is to be placed) with the knob of the jar. When the charge is positive, the ball may be made to recede from the wire three or four times by one charge, if it be replaced after each contact of the discharging rod; but when the charge is negative, it will recede but once: the cause of the ball's motion in these two instances is therefore different; in the latter it is attraction; in the former a continued current of electricity. In either of these experiments, if a strong charge is employed, the ball will be entirely driven out of the groove.

The perforation of a card, or of paper, by the electric explosion, has been also proposed as a test of the course of the electric fluid; but the effect of expansion interferes very much with its results. Two burs, or protrusions, are always produced; but Mr. Gough has lately shewn that when the experiment is made with accuracy, the bur on the positive side is constantly the smallest; and a hole made in card-paper by a punch exhibits a similar result, a small bur being raised on the side to which the punch is applied, and a larger bur on the opposite side.* I have been informed, that when a bullet is discharged through a sheet of copper analogous appearances are produced; and on making the experiment of electrical perforation on many bodies, less expansible than card, the indications were decidedly in favour of a current

* Nicholson's Journal; vol. xxxii. p. 176.

from the positive to the negative: such is particularly the case when a sheet of tinfoil or of thin lead is pierced; and it also occurs with thin pieces of wax or dry soap. These appearances are best observed with a magnifier after the explosions have been taken.

Mr. Symmer made the experiment of perforation with a paper book, in the middle of which he placed a sheet of tinfoil; on passing an explosion through them, the leaves were perforated, and the tinfoil indented in opposite directions; hence he concluded there was a double current, one fluid proceeding from the positive and the other from the negative.* If this experiment be attentively considered, it will be obvious that, by the interposition of the *tinfoil*, a *double interruption* in the *metallic* circuit is produced: now it has been already shewn, that at *every* such *interruption* a *spark* and an *expansive effect* invariably appear; consequently such an expansion must have occurred in the *paper* on *each side* of the *tinfoil*, and the perforation and burs prove that it did so: now the expansion of the paper in this way must

* Phil. Trans. vol. li. p. 371.

necessarily effect an indentation of the tinfoil opposite to it, and the indentations being on each side, from the paper to the tinfoil, must of course appear in opposite directions. Hence it is evident that Mr. Symmer mistook the expansive effects of the electric fluid for an indication of its direction; and his mistake has been ingeniously amplified by Mr. E. Walker, in a recent number of the Philosophical Magazine.*

Mr. Cavallo discovered that some mineral colours are affected by the passage of an electric charge over them; and this circumstance may be applied to shew the track of the fluid in passing from one side to the other of a card or thick paper, when the transmitting wires are at some distance from each other.

Experiment 66. Colour both sides of a card with vermilion, and place it upon the table of the universal discharger; one of the wires should be beneath the card, and the other in contact with its upper side; the distance of the points of the wires being one inch. If a charge be now passed through the wires, the fluid will pass from the positive wire across the surface of the

* Phil. Mag. vol. xlii. p. 161.

card to the part over the negative wire, and it will there perforate the card in its passage to the negative wire. The course of the fluid is permanently indicated by a neat black line on the card, reaching from the point of the positive wire to the hole; and by a diffused black mark on the opposite side of the card around the perforation, and next the negative wire. These effects are very constant, the black line always appearing on the side of the card which is in contact with the positive wire, and the perforation being near the negative wire.

I have lately contrived a means of demonstrating the direction of the electric fluid by its mechanical impulse, which confirms the general bearing of the preceding facts, and illustrates the assigned cause of their apparent anomalies. It has been long known, that a light float wheel, made by inserting several vanes of card-paper in the periphery of a cork that turns freely on a pin or center, will be put in motion by presenting it to an electrified point; and the motion of the wheel being always *from* the point, whether that was positive or negative, has been occasionally urged as an argument for a double current of electric fluid; although it is evident, from what has been previously stated, that a point, either positive or negative, must produce a *current* by the recession of the *air* opposed to it when *similarly* electrified by its contact; which is fully adequate to the production of these effects. Conjecturing that the currents of electrified air would not take place in this manner if the points were *opposed* to each other, I made the following arrangement.

Experiment 67. A light float wheel, of the preceding description, being mounted so as to turn *freely* between two upright wires, is placed on an *insulating* stem, and introduced between the pointed wires of the universal discharger, which are to be placed as accurately as possible opposite to each other, and at the distance of an inch or more from the upper vanes on their respective sides.—See Fig. 28. Now it is evident, from this disposition of the apparatus, that if there are *two* electric fluids moving in *opposite* directions, the wheel being *equally* acted on by each, will obey neither, and remain stationary; but if one only exists, it will receive motion in the direction that fluid passes. Connect one of the pointed wires with the positive conductor of an electrical machine, and the other with the negative conductor; as soon as the machine is turned the wheel will move, the direction of its motion being *from* the *positive* to the *negative* wire. Reverse the connections, so that the wire which was negative shall become positive, and that which was positive be rendered negative; the motion of the wheel will be *reversed,* for it will still turn *from* the *positive* to the *negative;* proving that the electric fluid actually moves in that direction. A similar effect will be produced by the discharge of a jar, provided it be *insulated* during the discharge, which is necessary to insure the transmission of the charge from one wire to the other, as it would otherwise be dispersed, by passing in various directions to the conducting bodies in contact with the outside of the jar.

Experiment 68. Place a card vertically, by inserting it in a small piece of cork that may form a base of about a quarter of an inch wide for it to stand on: the base should be barely sufficient to support the card in its vertical position, so that it may be overthrown by the slightest impulse. The pointed wires of the universal discharger being opposite to each other, and at about four inches distance, the card is to be stood upright on the table between them, and its height so adjusted that the line of direction between the wires may be about a quarter of an inch below its top edge. If the wires be now respectively rendered positive and negative, either by connecting them with the opposite conductors of the machine, or bringing them in the circuit of an insulated jar, the card will be thrown down, and constantly fall *from* the *positive to* the *negative*, changing its direction when the electrical connexions of the wires are changed.

If, instead of being placed midway between the wires, the card be put in *contact* with either of them, it falls *from* that wire, whether it be *positive* or *negative ;* but this arises from the card being an *imperfect conductor*, and consequently becoming *electrical* at its point of contact; as may be proved by covering it with silver leaf, which, by rendering it uniformly a *conductor*, prevents this effect. It will then *remain erect* if placed in *contact* with either wire, and *fall from the positive to the negative* if situated at equal distances from them.

Such are the leading mechanical phenomena of electricity, and such the indications they afford of the materiality of the electric fluid, and the nature of its diffusion and transmission. When the subtlety of this agent is considered, it must be admitted (I should presume) that we have few instances in philosophy where the action of an invisible power is so clearly defined by its effects; or where those effects tend so uniformly to one inference, as to render the phenomena intelligible by the mere exposition of their mutual relations and connexion.

CHAP. III.

Chemical Effects of Electricity.

THE agency of electricity, in the production of chemical changes, is even more remarkable and extensive than its mechanical power, and probably arises from the same cause. The most obvious and simple of them are connected with the appearance of light and the production of heat; and may therefore arise from the rapid motion of the electric fluid through the particles of other matter: for light and heat are also occasionally observed during the sudden mechanical compression of all elastic bodies. The simple motion of electricity through air is accompanied by a rise of temperature, as may be observed by introducing the bulb of a thermometer into the luminous current between two oppositely electrified balls of wood. The spark or explosion, and indeed every appearance of electric light, is accompanied by a peculiar smell, which has been considered as indicating that such appearances are the result of a species of combustion; but the continued appearance of the spark in a confined portion of air, in which it produces only a limited change: and its production under the surface of water and other fluids, militates strongly against this conclusion. In its concentrated state, electricity is capable of inflaming most combustible bodies, if passed as a spark through a stratum of air in contact with them; and any idea of the direct agency of heat, in the production of these effects, may be obviated by the transmitting conductor being formed of bodies that would absorb it.

Experiment 69. If ether, or highly rectified spirit of wine, be placed in a metal cup insulated and electrified, a spark may be drawn from the bottom of the cup through the spirit, by presenting to its surface either the finger, a brass ball, or even a piece of ice; and with any of these substances the spark will inflame the volatile fluid. If the spirit of wine be not highly rectified, it will be necessary to warm it moderately before the experiment; but this precaution is never necessary with ether.

Experiment 70. Dry the outside of a wine glass, that its stem may serve as an insulating stand; fill the glass nearly with cold water, and on the surface of the water pour a stratum of ether: connect the water, by means of a wire, with the conductor of the machine; when that is turned, if the knuckle be presented to the surface of the ether, a spark will pass from the water to the knuckle, and the ether will be set on fire.

The same effect will take place if a series of glasses, filled with a freezing mixture, and connected by wires, are employed to transmit the electricity from the machine to the water: so that it is evident the absorbing power of the intervening conductor does not prevent the power of the spark.

Experiment 71. Fill a flat porcelain dish with water, and on the surface of the water strew a quantity of powdered resin: place two wires at the opposite sides of the dish, with their ends near the surface of the water, and at four or five inches distant from each other: pass the charge of a jar from one wire to the other, and the resin in the track of the explosion will be inflamed. Similar effects are produced when the resin is strewed on the surface of a rough piece of wood or a loose ball of cotton.

Phosphorus placed in a little tin cup floating on water may be also readily inflamed, by passing a current of electrical sparks over its surface.

The most remarkable effects of combustion that are produced by electricity, result from its action on metals. Dr. Franklin was the first who observed these effects: his experiments were first extended by Mr. Kinnersly and by the celebrated Beccaria; and have since been pursued with great accuracy by Mr. Brook, Dr. Van Marum, and Mr. Cuthbertson.

Experiment 72. Place a strip of silver or gold leaf on white paper, and pass a strong shock through it: the metal will disappear with a bright flash, and the paper will be stained with a purple or grey colour.

Experiment 73. Take three pieces of window glass, each an inch wide and three inches long, place them together with two narrow strips of gold leaf between them, so that the middle piece of glass has a strip of gold on each of its sides; the extremities of the gold strips should project a little beyond the ends of the glass: pass the charge of a large jar through the gold strips; they will be melted and driven into the superfices of the glass. The outer slips of glass are usually broken, but that in the middle frequently remains entire, and is marked with an indelible metallic stain on each of its surfaces.

Experiment 74. The colours produced by the explosion of metals have been applied to impress letters or ornaments on silk and paper. The outline of the required figure is first traced on thick drawing paper, and afterwards cut out in the manner of stencil plates. The drawing paper is then placed on the silk or paper intended to be marked; a leaf of gold is laid upon it, and a card over that; the whole is then placed in a press, or under a weight, and a charge from a battery sent through the gold leaf. The stain is confined, by the interposition of the drawing paper, to the limit of the design; and in this way a profile, a flower, or any other outline figure may be very neatly impressed.

When a powerful electric charge is passed through a slender iron wire, the wire is ignited or dispersed in red hot balls. Very large batteries were formerly considered essential to the production of this effect; but if the wire be sufficiently thin, a single jar, exposing a coated surface of about 190 square inches, will sufficiently exemplify it. The finest flatted steel wire, sold at the watch-makers' tool-shops by the name of watch pendulum wire, answers exceedingly well. Cuthbertson's Balance Electrometer should always be employed to regulate the charge; the circuit from the inner to the outer surface of the jar should be as short as possible; and the wire intended to be melted placed in a straight line, and confined at the ends between small wire forceps.

Experiment 75. The inside of the Leyden jar, and the bent arm of the electrometer, being connected with the positive conductor of an electrical machine, and two inches of watch pendulum wire placed by means of the wire forceps between the lower insulated ball of the electrometer and the coating of the jar, the slider is to be set on the graduated arm of the electrometer to 15 grains. The machine is then to be put in motion, and when the intensity of the charge exceeds the resistance of 15 grains, the beam of the electrometer will descend, and the charge passing through the two inches of wire, will render it red hot, and melt it into balls.

Experiment 76. If the jar has not a paper

ring, it must now be breathed into, and eight inches of pendulum wire being placed in the circuit, the slider of the electrometer is to be set at thirty grains, and the turning of the machine resumed: when the charge is sufficiently intense, the beam of the electrometer will descend, and the charge passing through the eight inches of wire, will melt it with the same appearances as the two inches in the last experiment.

Experiment 77. Arrange eight inches of wire in the circuit, as in the last experiment; but instead of one jar charged to thirty grains, employ two jars charged to 15 grains. The wire will be melted precisely in the same manner; so that the effect is *equally* increased by *doubling* the *extent* of *coated surface*, or the *height* to which it is *charged.*

From numerous experiments of this kind, it has been concluded by Mr. Brook and by Mr. Cuthbertson, that the action of electricity on wires increases in the ratio of the square of the increased power; since *two* jars, charged to any degree, will melt *four* times the length of wire that is melted by *one jar;* and this will be again *quadrupled* by *doubling* the height of the charge.

This law, I have found, obtains in all accurate experiments with moderate lengths of wire, and it is apparent, in Mr. Cuthbertson's experiments, to some extent. The batteries of his construction usually contain fifteen jars, and one of these will just fuse half an inch of iron wire $\frac{1}{150}$th of an inch diameter; but the whole battery of fifteen jars will fuse sixty inches of the same wire.* I have made some experiments with very slender iron wire ($\frac{1}{250}$th of an inch diameter) on rather an extensive scale; but some of the charge is lost in pervading a considerable length of thin wire, and the explosion of the battery (at other times remarkably loud) is then scarcely audible. With a battery exposing forty feet of coated surface I have frequently melted eighteen feet of the last-mentioned wire by a single explosion, and the phenomena were remarkably brilliant, a shower of intensely ignited globules being dispersed in every direction.

The preceding law of the proportion of igniting power to the extent of coated surface, and height of charge employed, varies when any considerable difference exists in the *thickness* of

* Cuthbertson's Practical Electricity, p. 181, &c. or Nicholson's Journal, 4to. vol. ii. p. 525, &c.

the jars made use of; as thick jars display the same intensity with a comparatively small quantity of electricity, and consequently have less wire-melting power. I have in my possession a very large jar, which, from the extent of its coated surface, should melt three feet of wire with a charge of thirty grains; but from its limited electrical capacity in consequence of extreme thickness, it will melt only eighteen inches; and this is correspondent to the conclusion drawn by Mr. Cavendish, that the quantities of electricity required to charge different coated jars of the same extent will be in the inverse proportion of their thickness.*

The fusion of wire may therefore be employed as a measure of the quantity of electricity accumulated on any charged surface; for the preceding experiments shew that any given quantity of electricity will fuse the same length of wire, whether it be disposed on two jars or one; and hence it may be concluded, that the greater or less intensity of a charge does not materially affect its wire-melting power. This test is therefore practically useful; for the various electrometers measure only the intensity, and are

* Phil. Trans. vol. lxvi. p. 196, &c.

equally affected by one jar as by a battery of one hundred. When the fusion of wire is employed as a test of electrical power, care should be taken that the length of the circuit is always the same, and that the degrees of ignition are uniform; for a wire may be melted with but slight variations of appearance when very different quantities of electricity are passed through it. The lowest degree of perfect ignition ought therefore to be obtained in all comparative experiments, and its phenomena should be uniform; that is, as soon as the discharge is made, the wire should become red hot in its whole length, and then fall into balls.

The effects of gradually increasing the power of the charge, when wires of the same length and diameter are employed, are very remarkable. If the wire be iron or steel, its colour is first changed to yellow, then (by an increased charge) blue, by a further increase it becomes red hot, then red hot and fused into balls; if we continue to increase the charge, it becomes red hot and drops into balls, then disperses in a shower of balls, and lastly disappears with a bright flash, producing an apparent smoke, which, if collected, is a very fine powder, weighing more than the metal employed, and consisting of it and a portion of the oxygen of the atmosphere, with which it has combined.

The conversion of various metals into earth-like powders, of different colours, by exposing them to heat, with free access of air, has been long known; and modern chemists have accounted for such changes, by proving that a peculiar gaseous substance, which constitutes about a fifth part of our atmosphere, and is called oxygen, is constantly absorbed by metals when they lose their metallic appearance: and they have shewn that such changes do not occur by the mere agency of heat, unless air, or some other substance containing oxygen, be present. Hence the substances that result from the combustion of metals are called Oxides; such is the red lead of the shops, which is an oxide of lead; the crocus of the shops, which is an oxide of iron; and the grey powder used by lapidaries, which is an oxide of tin. These changes do not occur to all metals with the same facility: platina, gold, and silver undergo no change when exposed to the most intense heat; and are therefore usually converted into oxides by the agency of acids, which afford them oxygene more readily: but by the power of electricity all the known metals may be converted into oxides, and the circumstances of their change from the metallic state may be accurately investigated.

The most complete series of experiments that have yet been made on this subject were instituted by Mr. Cuthbertson.* The apparatus necessary for their demonstration consists of a glass cylinder, two or three inches diameter and eight inches high, mounted air tight with brass caps at each end; to the lower cap a stop-cock is screwed, and in the inside of the receiver, above the opening of the stop-cock, a small roller is fixed, on which a quantity of wire (attached to a packthread, that it may be readily moved forward) is coiled. A brass tube, about three inches long, is screwed into the centre of the upper cap, and through this tube the end of the packthread and wire is passed by means of a long needle; the tube is filled with hog's-lard secured by cork, so that the wire and packthread move through it air tight. The wire is by this means extended in the centre of the receiver,

* Nicholson's Journal, 4to. vol. v. p. 136, or Cuthbertson's Practical Electricity, p. 197.

and when one length is exploded another may be drawn forward by means of the packthread, to which the wire is attached at intervals of four inches; and thus many lengths of wire may be successively exploded without opening the receiver. To ascertain the quantity of air that has been absorbed during the process, a straight glass tube, about 10 inches long, may be screwed to the lower aperture of the stop-cock; the open end of the tube being then immersed in a vessel of quicksilver, and the stop-cock opened, the rise of the quicksilver in the tube becomes a measure of the absorption. This instrument is represented by Fig. 29, and its tube or gage by A.

The explosion of the wire alters the temperature of the air within the receiver considerably; it is therefore necessary, both before and after the explosion, to lay the instrument in a large vessel of water for some time, that the measure of the absorption may not be rendered inaccurate by the effect of expansion that attends an increased temperature. For this reason a narrow receiver is preferable to a wide one, for the temperature of the air within it is more easily restored; and the apparent anomalies that attended Mr. Cuthbertson's first experiments are thereby more effectually prevented.

If the air remaining in the receiver, after a sufficient number of explosions, be examined, it will be found to have lost a portion of its oxygene. And if, instead of atmospheric air, the receiver be filled with hydrogen or nitrogen, no oxidation of the metal will take place; but it will be fused, and very minutely divided.

The power of a large battery is necessary for the oxidation of metals, as to accomplish it completely, a higher power is required than that which is adequate merely to their fusion; the quantities of electricity required are not the same for every metal. Annexed is a statement of the comparative charges employed by Mr. Cuthbertson; the length of each wire exploded being ten inches. The column A expresses the diameter of the wire in parts of an inch. The column B the number of grains with which the electrometer was loaded; and C the colour of the oxide when collected in the receiver. The coated surface employed in all the experiments was the same; a battery of 15 jars, exposing about 17 feet of coated surface.

	A.	B.	C.
Lead wire	$\frac{1}{90}$	20	Light grey.
Tin wire	$\frac{1}{90}$	30	Nearly white.
Zinc wire	$\frac{1}{90}$	45	Nearly white.
Iron wire	$\frac{1}{150}$	35	Reddish brown.
Copper wire	$\frac{1}{150}$	35	Purple brown.
Platina wire	$\frac{1}{150}$	35	Black.
Silver wire	$\frac{1}{150}$	40	Black.
Gold wire	$\frac{1}{150}$	40	Brownish purple.

These experiments may be varied by exploding the wires, when stretched parallel to, and at about an eighth of an inch distant from the surface of a sheet of paper or glass; in either case a very beautiful figure is impressed, and on glass, a part of the metal in an unoxidated state appears immediately under the wire; whilst the oxidated portion produces a figure of some width by which it is encompassed. The colours of the oxides produced in this way, differ from those obtained in receivers, many colours being in some instances obtained from one metal.

The charges employed by Mr. Cuthbertson are rather high, and consequently attended with great risk of fracture to the jars of the battery; I have used in my experiments finer wires, and of shorter length, with a moderate charge. The proportions are indicated according to the preceding rule, in the following table.

The length of wire exploded in each experiment is five inches.

	A.	B.	*Colours of the figures on paper.*
Gold wire	$\frac{1}{180}$	18	Purple and brown.
Silver wire	$\frac{1}{160}$	18	Grey, brown, and green.
Platina wire	$\frac{1}{180}$	13	Grey and light brown.
Copper wire	$\frac{1}{160}$	12	Green, yellow, & brown.
Iron wire	$\frac{1}{180}$	12	Light brown.
Tin wire	$\frac{1}{180}$	11	Yellow and grey.
Zinc wire	$\frac{1}{180}$	17	Dark brown.
Lead wire	$\frac{1}{180}$	10	Brown and blue grey.
Brass wire	$\frac{1}{180}$	12	Purple and brown.

Brass wire is sometimes decomposed by the charge, the copper and zinc of which it is formed being separated from each other, and appearing in their distinct metallic colours when the explosion is made over a strip of glass. The figures of all the metallic oxides are usually more beautiful when impressed on glass, than paper, but their colours are less permanent.

The oxides produced by electrical discharges appear to consist of several distinct portions, of different degrees of fineness; when a wire is exploded in a receiver, part of the oxide immediately falls to the bottom; but another portion remains suspended in the air for a considerable time, and is at length gradually deposited. It is probable this circumstance may partly occasion the different colour of oxides produced in close receivers and in the open atmosphere, for in this last a portion of the oxide is always lost.

The chemical agency of electricity is the more remarkable, since it tends equally to promote combination, or decomposition. Metallic oxides already formed, may be restored to the metallic state by its means; and for this purpose a very simple arrangement will suffice.

Experiment 78. Introduce some oxide of tin into a glass tube, so that when the tube is laid horizontal the oxide may cover about half an inch of its lower internal surface. Place the tube on the table of the universal discharger, and introduce the pointed wires into its opposite ends, that the portion of oxide may lay between them. Pass several strong charges in succession through the tube, replacing the oxide in its situation, should it be dispersed. If the charges are sufficiently powerful, a part of the tube will soon be stained with metallic tin, which has been revived by the action of the transmitted electricity.

Other metallic oxides may be revived in the same way; or if vermilion be employed, which consists of sulphur and mercury, the mercury will be separated, and that with such facility, that the charge of a very moderate sized jar is fully sufficient.

When the electric spark is taken in various fluids, they are decomposed by it. Water is converted into two gases, oxygen and hydrogen; in the proportion of two measures of the latter to one of the former; and when a sufficient quantity of these are liberated, if an electric spark be sent through them, they inflame and disappear, water being reproduced.

This experiment was first made by a society of Dutch chemists, assisted by Mr. Cuthbertson; it is a very laborious and tedious one,* but has been much facilitated by an arrangement of

*. See Dr. Pearson's paper in Nicholson's Journal, 4to. vol. i. p. 241, &c. or Phil. Trans. vol. lxxxvii. p. 142.

Dr. Wollaston's.† Two very fine wires of gold or platina are inserted in capillary tubes; one extremity of each wire is previously pointed as fine as possible, and being introduced within the capillary tube to a short distance from its end, the glass is softened by heat until it adheres to the point and covers it: the glass is then gradually ground away until the point of the gold wire is perceptible through a lens. Two wires so prepared are introduced into a vessel of water, so that the points may be near each other, and form an interrupted metallic circuit: one of the wires is to be connected with the ground, or with the negative conductor of the machine; and the other with an insulated ball placed at a short distance from the positive conductor. When a current of sparks are passed in this way, a series of minute bubbles of gas rise from the points of the gold wires, which may be collected in a small inverted receiver placed over the ends of the wires, and will explode on the application of a lighted taper, or by passing an electric spark through it: but it is to be observed, that a considerable time is required to produce a sufficient

† Phil. Trans. vol. xci. p. 427.

quantity for this purpose. Dr. Wollaston found that the rapidity of the decomposition was in proportion to the limited size of the point exposed; a point $\frac{1}{700}$ of an inch diameter effected the decomposition, when the spark from the conductor to the insulated ball was 1-8th of an inch long; and a point $\frac{1}{1500}$ of an inch diameter, produced the same effect with sparks $\frac{1}{20}$ of an inch in length. These coated wires are necessary for the decomposition of all conducting fluids; as they confine the action of the electricity to a single point, and prevent the diminution of intensity that would otherwise ensue.

With less perfect conductors, as oils, alcohol, and ether, this precaution is unnecessary; they may be decomposed in the apparatus, Fig. 27; sparks being passed from the platina wire to the bottom of the metal dish without any risk of fracturing the tube, and the gaseous product rising into the tube, which serves as a receiver. The gases procured from inflammable fluids are chiefly what are called hydro-carbonates, and consist of hydrogen, holding in solution various proportions of charcoal. When concentrated acids are decomposed, the gaseous product is usually oxygen gas.

Dr. Wollaston transmitted a current of electricity, by means of two fine gold points, along the surface of a moistened card tinged with litmus; after a few turns of the machine a redness was perceptible about the positive wire. The negative wire being afterwards placed on this blue spot, soon restored it to its original blue colour. It therefore appears that the effect of an acid is by some means produced at the positive wire, and that this effect is counteracted by reversing its electricity.

The same philosopher inserted two silver wires (coated with sealing-wax, so that their ends only were exposed) into a solution of copper: on transmitting a current of electricity from one wire to the other, the receiving or negative wire became coated with copper; and the copper coating was removed when the electricity of the wire was reversed.

The facility with which the electrical spark can be taken in air-tight vessels, renders its application to chemical purposes an object of the first importance; and in no instance is its action more extensive than in its application to the combination or decomposition of various gases. The formation of water is shewn by filling the tube, Figure 24, with quicksilver, and inverting it in a vessel of the same; a mixture of hydrogen and oxygen gases, in the proper proportions, are then to be passed up into the tube, until they occupy about an inch of its upper part. A spark being now passed between the wires, the gases will inflame, and the quicksilver will rise to the top of the tube covered with a thin film of water, which has resulted from the combustion of the gases. For all experiments of this kind the glass vessels should be at least half an inch thick, that they may resist the expansion produced in the gases by the explosion.

A perceptible quantity of water may be formed in this manner by employing a stout globe, with a stop-cock; a wire passing through its centre to within a short distance of the cap to which the stop-cock is screwed. The globe is to be exhausted by means of an air-pump, and then screwed on a receiver containing a mixture of oxygen and hydrogen gases, and furnished with a stop-cock.—See Figure 30. The cocks being opened, the globe will be filled with the gases; they are then to be shut, and a spark passed from the wire in the inside of the globe to the cap. A bright flash ensues, and the inside of the globe becomes coated with moisture; the cocks are then to be opened, and more gas will rush into the globe. The cocks being again closed, a second explosion may be made, which will increase the dew on the inside of the globe; and the experiment may be repeated in this way until drops of water are perceived.*

Experiment 79. Take a stout glass tube closed at one end, and having two wires passed through its sides so as to admit of a spark being taken within the tube. Adapt a cork to the open end of the tube, and, holding it inverted, pass into it a mixture of hydrogen and oxygen, or of hydrogen and atmospheric air. Close the tube with its cork, and pass a spark through it; a loud explosion will follow, and the cork will be expelled with violence. An apparatus is sometimes fitted up for this purpose as an amusive experiment; it is called the inflammable air pistol.

The facility with which inflammable air is lighted by even a moderate electric spark, induced professor Volta to contrive his inflamma-

* This experiment was first made by Mr. Cavendish, in the year 1781.

ble air-lamp; (for a modification of which, a patent was some time since obtained as a source of instantaneous light.) It consists of a reservoir filled with hydrogen gas, subject to the constant pressure of a column of water, and confined by a stop-cock, which, when opened, permits it to escape in a slender stream from a small aperture. In a box beneath the vessel of gas an electrophorus is placed, and a wire passes through a glass tube from the upper part of this box to the opening of the stop-cock. The cover of the electrophorus is connected by a silk string with the handle of the stop-cock; so that the same motion that opens the cock, raises the cover of the electrophorus, and the spark that strikes from it, is conveyed by the insulated wire to the stream of gas, which it inflames. This effect takes place every time the stop-cock is opened, for the electrophorus will produce sparks for a considerable time, without any new excitation; and the quantity of gas consumed at each repetition of the process is inconsiderable, so that a light may be procured above a hundred times before the contents of the reservoir is expended; and it may then be easily replenished.

Dr. Priestley observed, that when electrical sparks were taken for a considerable time in a confined portion of common air, the bulk of the air was diminished; and on introducing any blue vegetable liquor into the vessel in which the experiment had been made, a redness resulted, evincing the presence of an acid. Mr. Cavendish repeated this experiment with great precision, and proved that the elements of atmospheric air (oxygen and nitrogen,) were by this means combined in a different proportion, and formed nitric acid.* The air on which the experiment was made, was confined between two columns of quicksilver in the angular part of a bent tube, first filled with quicksilver, and then inverted with its legs in two separate glasses of the same fluid; (see figure 31;) the air being introduced into the tube after its inversion, so as to occupy the upper part of the angle for the length of an inch, or an inch and a half. The quicksilver in one of the glasses was then connected either with the negative conductor or the ground, and that in the other with an insulated ball placed near the po-

* Phil. Trans. vol. lxxv. p. 372, and vol. lxxviii. p. 261.

sitive conductor, and in this way a current of sparks was passed through the confined portion of air; and as the bulk of this diminished, fresh portions were passed up into the tube; so that the length of the column was preserved nearly equal. The experiment was rarely completed in less than a fortnight or three weeks, the machine being worked for half an hour each day. The internal diameter of the tube should be about one-tenth of an inch: instead of the form above described, it may be straight, with a platina wire sealed into one end, and its open extremity immersed in a basin of quicksilver. (Figure 32.)

Mr. Cavendish found that the experiment succeeded best when instead of common air alone, a mixture of five parts of oxygen and three parts of common air were employed: such a mixture disappears almost entirely; and if a small quantity of soap lees, or solution of potash, be passed up into the tube, the process is accelerated, and the solution becomes nitrate of potash, or saltpetre.

This experiment was twice repeated on an extensive scale by Mr. Gilpin, under the direction of Mr. Cavendish and other members of the Royal Society. There was some slight difference in the proportions of the gases absorbed, but from the mean result of the experiments, it appears that seven measures of oxygen unite to three measures nearly of nitrogen, to form nitric acid.

The experiments that may be made on the combination and decomposition of various gases are very numerous; to describe them in detail would be useless to those who are not acquainted with the leading facts of chemistry; and the object of the chemical student will be as effectually accomplished by a tabular enumeration of the gases that are affected by electricity, and the results they afford. The gases are usually exposed to the action of electricity in a closed tube, with two wires passing through its sides near the closed end: the tube is filled with mercury, and inverted in a vessel of the same; and the gas being then introduced until it presses the mercury below the wires, sparks are passed between them until the required change is produced; with mixtures of inflammable gases and oxygen the first spark usually produces the change, but with other mixtures it is sometimes necessary to continue the current of sparks for hours.

When figures are prefixed to the gas, or its result, as stated in the following table, they indicate the proportional measures employed or produced; they are introduced principally in cases where the use of different proportions occasions a variation of result.

Mixed Gases.	*Result.*
Atmospheric air and hydrogen ..	Water and nitrogen.
Oxygen and hydrogen	Water.
Chlorine and hydrogen	Muriatic acid.
Muriatic acid and oxygen	Chlorine.
Carbonic oxide and oxygen	Carbonic acid.
Nitrogen and oxygen	Nitric acid.
Sulphurous acid and oxygen	Sulphuric acid.
Phosphuretted hydrogen and oxygen	Water and phosphoric acid.
Sulphuretted hydrogen and oxygen	Water and sulphurous acid.
Oxygen and ammonia	Water and nitrogen.*
100 olefiant gas and 284 oxygen	Carbonic acid and water.
100 olefiant gas and 100 oxygen	Carbonic oxide and hydrogen.
100 carburetted hydrogen and 100 oxygen	Carbonic oxide and hydrogen.
100 carburetted hydrogen and 200 oxygen	Carbonic acid.

* If there be an excess of oxygen, nitric acid is also a product.

Compound Gases.	*Result.*
Muriatic acid	Hydrogen.*
Fluoric acid	Hydrogen.*
Nitrous gas	Nitric acid and nitrogen.
Carbonic acid	Carbonic oxide and oxygen.
Sulphuretted hydrogen	Sulphur and hydrogen.
Phosphuretted hydrogen	Phosphorus and hydrogen.
Ammonia	Hydrogen and nitrogen.
Olefiant gas	Charcoal and hydrogen.
Carburetted hydrogen	Charcoal and hydrogen.

And, from analogy, it is probable that all compounds of Hydrogen with inflammable matter are equally susceptible of electrical decomposition.

These various effects, produced by the same agency, do not appear susceptible of any other explanation than that which assumes the action of electricity to be mechanical; and even on this assumption they are not strictly intelligible. The momentary agitation into which the various mediums are thrown by the action of the spark, might be considered as likely to promote a new arrangement of parts; but, admitting this, why is the change instantaneous in some instances, and gradual in others? and by what inversion of principle is the same impulse that unites the

* On the authority of Dr. Henry and Mr. Dalton.

particles of bodies, enabled subsequently to separate them? These are questions it would be interesting to resolve; but there appears no clue by which such intricate processes can be at present analysed. The chemist must therefore be content to avail himself of the practical advantages they afford to his art, and await the progress of discovery for the development of their theoretical relations.

The luminous phenomena of electricity sufficiently prove the influence of the electric fluid on light, and this fact is remarkably confirmed by its agency in the production of phosphoric appearances in various bodies when it passes luminously over their surface. Experiments of this kind were first made by Mr. Lane and Mr. Canton;* and have been since extended by Wilson, Morgan, and Skrimshire.†

Experiment 80. Place a piece of dry chalk on the table of the universal discharger, and adjust the wires on its surface, with their ends at two inches distant from each other. Pass a strong charge from wire to wire, and after the

* Priestley's History of Electricity, p. 312.

† Nicholson's Journal, vol. xv. p. 281; vol. xvi. p. 101; vol. xix. p. 153.

explosion a streak of light will be evident in the track of the discharge, exhibiting the prismatic colours: it will continue for some seconds.

Similar effects ensue when the charge is passed over the surface of various other bodies; but the colour and the duration of the light vary considerably; and if the charge is passed through the substance of some of them, they will be dispersed in luminous particles that retain their light for a considerable time. The following table contains an enumeration of several substances that may be thus rendered phosphorescent, and displays the results they produce.

Native sulphate of barytes	Bright green light.
Native carbonate of barytes	Ditto, less brilliant.*
Acetate of potash (dry)	Brilliant green light.
Succinic acid	Ditto, and more durable.
Loaf sugar.	Ditto.
Specular gyp m, or selenite	Ditto, but transient.
Calcined oyster shells	Prismatic colours.
Ditto calcined with sulphur	Durable and bright light.
Rock crystal	Light first, red, and then white.
Quartz	Dull white light.
Borax	Faint green light.
Boracic acid	Bright green light.

* These results are taken from experiments made with the specimens I chanced to possess.

The bodies here enumerated form but a very small proportion of those that become phosphoric by the discharge, but they are such as possess that property in the most remarkable degree. For a systematic examination of a very extensive series of substances the reader is referred to Mr. Skrimshire's papers in Nicholson's Journal.

Equally remarkable with this property of exciting phosphoric phenomena, is the effect of an electric explosion on various opaque bodies: it was first observed by Dr. Priestley, and may be thus exhibited:—

Experiment 81. Let two wires be fitted into a groove on the surface of a piece of smooth mahogany, so that, by sliding the wires backwards or forwards, their ends may be placed at any required distance from each other. When they are about half an inch apart, place a thumb or finger over the interval, and pass a charge from wire to wire; the thumb will appear perfectly transparent during the passage of the spark beneath it, but no unpleasant sensation will be felt.

Experiment 82. Substitute a jar of water, or any coloured fluid, in the place of the thumb:

when the discharge is made, the fluid will be distinctly and curiously illuminated.

Experiment 83. Place the ends of the wires at the distance of three-fourths of an inch; and over the interval lay a thick piece of pipe-clay or of pumice-stone: when the charge passes, these opaque substances will appear perfectly transparent.

Experiment 84. Arrange five or six eggs in a straight line, and in contact with each other: pass a small shock through them, and they will seem perfectly luminous.

Experiment 85. Insert two wires, so as to come within a short distance of each other, in a small melon, an orange, or apple: pass a shock through the wires, and the fruit will appear transparent.

These experiments are susceptible of considerable variety, since every substance that is not a good conductor, becomes more or less luminous by the passage of the charge: but no correspondence has been yet observed between the existence of this property and the chemical characters of the substances in which it obtains.

It may be proper here to mention, that for experiments on electric light, and the excitation of phophorescence, it is absolutely necessary to operate in a dark room; for the presence of the least extraneous light will prevent the observation of such phenomena.

In addition to the mechanical and chemical agencies of electricity already described, it has been observed to have some relation to the phenomena of magnetism. The needle of the ordinary ship compass has been often noticed to vary during a thunder storm, and in some instances its poles have been reversed. Dr. Franklin passed the charge of some large electrical jars through fine steel needles; their ends were blued, and magnetic polarity communicated to them. The effect depends principally on the situation of the needles when they are struck, and is little affected by the manner in which the charge is passed through them; the communicated magnetism is strongest when the needle is struck, whilst it lies in a direction from north to south; and weakest when it points from east to west.

Experiments of this kind are most effectually made with needles of steel wire, the fortieth or fiftieth of an inch diameter, and three or four inches long, with a small dent in the middle by which they may be supported on a point: the charge of a battery should be passed through them.

Experiment 86. Place a steel wire of the preceeding description in the direction from north to south, and pass a moderately strong charge of a battery through it; it will become magnetic, the end that lay southward being the south pole.

Experiment 87. Render a steel wire slightly magnetic, and place it in the magnetic meridian, with its south pole towards the north. A strong charge of a battery will either destroy its magnetism, or reverse its magnetic poles; if its magnetism is merely destroyed, a second charge will magnetize it anew, but with reversed poles.

Experiment 88. Place a steel wire in a perpendicular direction, and pass a strong charge through it; it will become magnetic, the upper end being the south pole. If this end be now placed downwards, the transmission of another charge will destroy its magnetism, or reverse its poles.

A strong charge passed through a natural magnet destroys its power.

These phenomena convey no information relative to the nature of the electric fluid; nor can they justly be considered as displaying any analogy between it and the cause of magnetism; for that power is also excited and modified by mechanical action, and by the agency of heat; and it is probable that in this particular instance, the three influential causes operate nearly on the same principle: but until the nature of magnetism is more clearly developed, it is useless to speculate on the probable action of other powers in exciting or modifying its phenomena.

PART III.

NATURAL AGENCIES OF ELECTRICITY.

CHAP. I.

On the Identity of Electricity, and the Cause of Lightning.

During the consideration of the nature and peculiarities of electrical action, the principal stimulus to inquiry must arise from the novelty and singularity of the appearances observed, and some difficulty may be experienced in fixing the attention on a series of facts apparently insulated and unconnected with the usual sources of human interest: but in the experimental sciences these preliminary steps are always necessary, and may be considered analogous to that collection and preparation of rude materials which is the essential precursor of every useful and valuable production of art.

The utility of electrical science is most evident in its application to the phenomena of nature; and more remarkably so, when the vague hypotheses and conjectures that were previously applied are compared with our present conception of the subjects it has explained.

The philosophy of the ancients appears to have been remarkably defective in its application to the phenomena of the atmosphere; many of the most interesting effects were entirely unobserved, and the production of its luminous phenomena was described by their best naturalists as amongst "the awful mysteries of Nature;" whilst the phenomena themselves were invested by their poets with the character of instruments of punishment and revenge in the hands of their deities.

In modern times, before the discoveries in electricity, these phenomena were referred to the inflammation of subtle effluvia or sulphureous exhalations in the higher regions of the atmosphere, or to the contending percussions of one cloud against another! In this way is the place of true knowledge supplied for ages by dreams and fables; so disposed is the human mind to create imaginary causes, when those that really operate are beyond our comprehension.

In the infancy of this science, Mr. Grey, (who had increased the effects of electrical excitation by the application of his discovery of conductors and non-conductors to the improvement of the apparatus,) was led, by considering the appearance of electricity, when passing from one conductor to another, to notice a faint similarity between the snap and light of the spark, and the phenomena of thunder and lightning. His remark does not appear to have attracted the attention of electricians, until, in 1748, the Abbe Nollet published, in the fourth volume of his "Lecons de Physique," the following extension of this conjecture: "If any one should take upon him to prove, from a well-connected comparison of phenomena, that thunder is in the hands of nature, what electricity is in ours, that the wonders we now exhibit at pleasure, are little imitations of those great effects which frighten us; and that the whole depends on the same mechanism: if it is to be shewn, that a cloud prepared by the action of the winds, by heat, by a mixture of exhalations, &c. is when opposite to a terrestrial object, that which an electrified body is, when at a certain distance from one that is not electrified; I confess this idea, if it was well supported, would afford me much pleasure; and to support it, how many specious reasons present themselves to a man well versed in electricity. The universality of the electric matter, the rapidity of its action, its inflammability, and its activity in inflaming other bodies; its property of striking bodies externally and internally, even to their smallest parts, the remarkable example we have of this effect in the Leyden experiment, the idea we may legitimately form in supposing a greater degree of electric power, &c. All these points of analogy, which I have for some time meditated, begin to make me believe that one might, by taking electricity for a model, form to one's self, in relation to thunder and lightning, more perfect and probable ideas than any that have been offered hitherto."*

This remarkable observation reflects considerable credit on the Abbé Nollet for his penetration and sagacity; but it bears no comparison with the acute conception, sound philoso-

* Lecons de Physique Experimentale, tom. iv. p. 314.

phical argument, and satisfactory experiments, by which Dr. Franklin has demonstrated the identity of the electric fluid, and the cause of thunder. This excellent philosopher made very numerous original Observations on the phenomena of electricity, which were communicated in a Series of Letters to a member of the Royal Society, from the year 1747 to 1753. In these, amongst an unexampled variety of electrical discoveries, he first detailed an hypothesis to explain the phenomena of thunder storms by the known properties of electricity, and afterwards demonstrated the truth of his supposition by the most extraordinary experiment ever made. Dr. Franklin had observed with equal attention the peculiarities of the natural phenomenon, and the power to which he ascribed its production; he enumerated the following, as their leading features of resemblance.

1st. The zigzag form of lightning corresponds exactly in appearance with a powerful electric spark that passes through a considerable interval of air.

2d. Lightning most frequently strikes such bodies as are high and prominent, as the summits of hills, the masts of ships, high trees, towers, spires, &c. The electric fluid, when striking from one body to another, always passes through the most prominent parts.

3d. Lightning is observed to strike most frequently into those substances that are good conductors of electricity, such as metals, water, and moist substances; and to avoid those that are nonconductors.

4th. Lightning inflames combustible bodies. The same is effected by electricity.

5th. Metals are melted by a powerful charge of electricity. This phenomenon is one of the most common effects of a stroke of lightning.

6th. The same may be observed of the fracture of brittle bodies, and of other expansive effects common to both causes.

7th. Lightning has been known to strike people blind. Dr. Franklin found, that the same effect is produced on animals when they are subjected to a strong electric charge.

8th. Lightning destroys animal life. Dr. Franklin killed turkies of about ten pounds weight, by a powerful electric shock.

9th. The magnetic needle is affected in the same manner by lightning and by electricity, and iron may be rendered magnetic by both causes.

The phenomena are therefore strictly analogous, and differ only in degree; but if an electrified gun-barrel will give a spark, and produce a loud report at two inches distance, what effect may not be expected from perhaps 10,000 acres of electrified cloud? And is not the different *extent* of these conductors, equal to the different limit of their effects?

But, to ascertain the accuracy of these ideas, let us have recourse to experiment. Pointed bodies receive and transmit electricity with facility; let therefore a pointed metal rod be elevated in the atmosphere, and insulated; if lightning is caused by the electricity of the clouds, such an insulated rod will be electrified whenever a cloud passes over it, and this electricity may then be compared with that obtained in our experiments.

Such were the suggestions of this admirable philosopher: they soon excited the attention of the electricians of Europe, and having attracted the notice of the King of France, the approbation he expressed excited in several members of the French Academy* a desire to perform

* Messrs. Dalibard, De Lor, Mazeas, Buffon, and Monnier.

the experiment proposed by Franklin, and several insulated and pointed metallic rods were erected for that purpose. On the 10th of May 1752, one of these, a bar of iron 40 feet high, situated in a garden at Marly, became electrified during the passage of a stormy cloud over it; and during a quarter of an hour it afforded sparks, by which phials were charged, and other electrical experiments performed. During the passage of the cloud a loud clap of thunder was heard, so that the identity of these phenomena was thus completely proved. Similar experiments were afterwards made by M. de Lor, Buffon, and Monnier, in France, and by Dr. Watson, Mr. Canton, Mr. Wilson, and Dr. Bevis in England.

Dr. Franklin had not heard of these experiments, and was waiting the erection of a spire at Philadelphia to admit an opportunity of sufficient elevation for his insulated rod, when it occurred to him that a kite would obtain more ready access to the regions of thunder than any elevated building. He accordingly adjusted a silk handkerchief to two light strips of cedar placed crosswise; and having thus formed a kite, with a tail and loop, at the approach of the first storm he repaired to a field accompanied by his son.

Having launched his kite with a pointed wire fixed to it, he waited its elevation to a proper height, and then fastened a key to the end of the hempen cord, and attached this by means of a silk lace (which served to insulate the whole apparatus) to a post. The first signs of electricity which he perceived was the separation of the loose fibres of the hempen cord: a dense cloud passed over the apparatus, and some rain falling the string of the kite became wet; the electricity was then collected by it more copiously, and a knuckle being presented to the key, a stream of acute and brilliant sparks was obtained. With these sparks spirits were fired, phials charged, and the usual electrical experiments performed: and thus was this important discovery, which its author had modestly called an hypothesis, established as a scientific truth.

Dr. Franklin afterwards constructed an apparatus for perpetual observations; it consisted of an insulated rod placed on the top of his house, and connected with two bells and a pendulum, which were so arranged as to ring when electrified, and thus gave notice of the approach of a charged cloud.

These experiments were repeated in almost every civilized country, and with various success: in France a most formidable result was obtained by M. de Romas; he had constructed a kite of seven feet in height and three feet wide, which was raised to the height of 550 feet, by a string having a wire interwoven through its whole length to render it a better conductor: from the string of this kite, on the 26th of August 1756, sparks, or rather streams, of lightning were darted to the earth, of an inch in diameter and ten feet long, whilst the preliminary phenomena were equally terrific.*

It would have been surprising had such experiments been constantly conducted with security: in the management of the ordinary electrical apparatus shocks are often received inadvertently, and the first experimenters on atmospherical electricity were often severely shaken. It is indeed rather fortunate that, amidst the thousands of experiments of this hazardous nature that have been made since the first dis-

* Memoires des Savans Etrangers, tom. ii. p. 393, & tom. iv. p. 514, &c.

covery, but one fatal catastrophe has occurred, and that happened at so early a period as to preclude the chance of proper precaution having been employed. The individual, who will be immortalized as the victim of electrical science, was Professor Richman of Petersburgh. He had constructed an apparatus for observations on atmospherical electricity, which was entirely insulated, and had no contrivance for discharging it when electrified too strongly. On the 6th of August 1753, he was examining the electricity of this apparatus, in company with a friend: whilst attending to an experiment his head accidentally approached the insulated rod, and a flash of lightning immediately passed from it, through his body, and deprived him of life. A red spot was produced on his forehead, his shoe was burst open, and part of his waistcoat singed; his companion was benumbed, and rendered senseless for some time; the door-case of the room was split, and the door torn off its hinges.*

This accident demonstrated the necessity of the greatest caution in such experiments;

* Phil. Trans. vol. xlviii. p. 765; or Priestley's History, p. 358.

and they are now usually rendered secure by placing a metal rod, connected with the ground or the nearest water, at a small distance from the end of the insulated rod; with this arrangement, when the electricity becomes too powerful, it passes the interval between the rods, and is conducted safely to the ground.

These experiments sufficiently prove the agency of the electric fluid in the production of lightning and thunder. Dr. Franklin had anticipated, as a consequence of the verification of his conjecture, that the known properties of electricity might be employed to provide some security against the concomitant dangers of a thunder storm; for if the injury usually sustained arises from the discharge of a large quantity of electric fluid, it may be prevented by providing a proper conducting channel for such discharge. Metals conduct electricity better than any other substances known, and when buildings or ships are struck by lightning the greatest damage is usually effected between detached pieces of metal, since the electric fluid, in passing through the intervals, produces the usual expansive effect observed in every interrupted metallic circuit. Hence Dr. Franklin proposed to erect a perfectly continuous metallic rod by the side of any building intended to be secured from the effects of lightning: the rod was to be pointed at each extremity, and extended above the highest part of the building and below its foundation, the lower extremity being connected with the nearest water or other conducting matter. In this way a perfect metallic circuit is provided, and through this any electrical discharge would pass more readily than through the detached or imperfect conductors of which the building is composed. Experience has sufficiently demonstrated the utility of this proposed method of defence, and it is now very generally adopted in situations where the chance of injury from thunder storms is considerable. There are some experiments usually employed to illustrate its application, and these it may be proper to notice in this place.

Experiment 89. Construct a pyramid of several pieces of wood placed one on the other; let the lowest piece form the square base of the pyramid, and the upper conical part rest upon the base by three small brass balls. A square hole is to be made in one side of the base, and a piece of wood fitted loosely in it; upon this loose piece of wood one of the brass balls is to rest, so that the upper part of the pyramid can only preserve its equilibrium, whilst this wood is in its situation. Pieces of wire are let into the several parts of the pyramid in such a manner, that when it is put together the ends of the wires may come in contact, and form a continuous conductor from the top to the bottom: the part of this conductor which traverses the loose piece of wood, may, by reversing that, be removed from its continuity with the others, and will consequently occasion an interruption in the metallic circuit. If a charge is passed through the conductor when this interruption exists, the piece of wood will be thrown out, and the pyramid will fall; but if the piece of wood is placed so as to render the conductor continuous, the charge will be conducted without producing any effect.

In the same way, a model of a powder mill may be blown up, or a house be set on fire, by making an interrupted circuit within them, and placing gunpowder, or other combustible matter in its interval. These models are usually furnished with a moveable conductor; when this is affixed to them, the charge of a large jar or battery will produce no effect; but when the conductor is removed, and the charge passes through the interrupted circuit, the combustibles are inflamed, and devastation ensues. Apparatus properly fitted up for these demonstrations are sold by the philosophical instrument makers.

Experiment 90. To exemplify the method of defending ships, a small model may be fitted up, with a glass tube for the mast, to which all the rigging should be attached: the tube is to have two wires inserted through its opposite ends, until within half an inch of each other; it is then to be filled up with water and the ends stopped. The lower wire is to be connected with a small metal thread tied to the stern; the upper wire is to be surmounted by a small ball, which may serve as the top of the mast. A moveable conductor may be formed of a thin copper wire placed parallel with the mast, rising above it, and connected at bottom with the metal thread. If a powerful charge is passed along the mast whilst the conductor is attached to it, no effect is produced; but if the conductor is removed, the charge passes through the body of the mast, and shatters it to pieces.

If the conductors employed in these experiments are pointed at the extremity, it will be found difficult to make the charge pass with an explosion, and it will in general be transmitted either silently or with considerably diminished force. It was on this principle that Dr. Franklin recommended that the ends of the conductors, erected to preserve buildings from the effects of lightning, should be acutely pointed; for a point will always reduce the intensity of any electrified surface to which it is presented, and will act upon its electricity at a considerable distance: a pointed conductor must consequently tend to weaken the charge of any cloud that may approach it, before it comes near enough to explode; and if the approach of the cloud be gradual, its charge will be even entirely dissipated without explosion. These facts may be readily exemplified by experiment.

Experiment 91. Suspend a number of downy feathers, or an expanded lock of cotton from the conductor of an electrical machine, so as to represent a cloud. When the machine is turned, the fibres will expand and separate from each other; and if a large ball or a flat surface be presented, they will extend themselves towards it: but if the ball be removed, and a pointed wire be presented at the same distance, the divergence will diminish, and the fibres will collapse and shrink up from the point.

Experiment 92. Charge a very large jar; touch the outer coating with one hand, and bring a sharp needle held in the other, very gradually toward the charging wire; the jar will be entirely discharged without explosion. Repeat the experiment, approximating the needle more rapidly. A very inconsiderable explosion will be produced.

If the point used in these experiments be not connected with the ground or the outside of the jar, by a perfectly continuous conductor, it will not produce a silent discharge, but will receive an explosion like a ball. Hence the necessity, that conducting rods be perfectly continuous, and well connected with some considerable mass of conducting matter, as water, water pipes, or a moist stratum of earth.

At one time there was a ridiculous dispute amongst electricians concerning the propriety of pointed terminations to conductors, and it was proposed by some to place a ball of moderate size on them, to prevent their action as points. Independent of the weakness of the arguments employed to shew that on some occasions a blunted conductor might be preferable to a pointed one, the size of the balls proposed was such, that when opposed to a thunder-cloud, or even to a powerful electrical machine, they would have virtually acted as points!

There are two instances on record, of houses in this country having been struck by lightning, so as to sustain some damage, although they were furnished with pointed conductors. The conductors attached to these buildings were not constructed in the most perfect manner, and had they been so, the occurrence of two failures in so many years, and amongst the thousands of conductors that have been erected, is a circumstance scarcely to be urged as an objection.

This method of defence can scarcely fail to be effectual, if employed with an attention to the following circumstances. The conducting rod, or rods, (for if the building is large there should be several,) should be formed of copper or iron, three quarters of an inch thick. Its upper extremity should be acutely pointed, and rise three or four feet above the highest part of the building. The parts of which the rod is formed should be joined closely, the ends that are applied to each other being screwed together. All the metallic parts of the roof should be connected with the rod, and it should pass down in as direct a line as possible, and penetrate several feet below the foundation, from which it should be inclined outwards. The underground part of the conductor is better formed of copper, to prevent its decay; it should be connected, if it possibly can, with a moist stratum of earth, or with a large body of water. The penetration of the conductor to some depth below the level of the foundation, will in many instances procure this advantage for it.

The conductor is sometimes made wholly of copper, it may then be thinner than if made of iron; for a stationary conductor, I should conceive, that a copper rod of half an inch thick would answer every required purpose; and there is little doubt that a less quantity of metal made into a hollow tube, so as to increase its superficies, would be equally, or even more effectual.

Conductors for ships have been made of

chains, (which are highly improper,) and of copper wires, which are easily attached; but they are with equal ease detached; and I have been informed by several captains, that many ships furnished with such conductors keep them in an inactive state, packed up below, during long and hazardous voyages For this reason it would be better that fixed conductors should be employed; they might, I should conceive, be attached to the mast; and where motion is required, an interruption might be made in the inflexible conductor, and its parts be connected together by a length of spiral wire, which would be at once perfectly continuous, and sufficiently flexible to yield to every necessary movement.

Although the various metals are affected very differently by the same quantity of electricity discharged through them, yet they are all proper for the construction of conductors, provided they are employed of sufficient thickness to resist fusion by a stroke of lightning: for this reason the gutters, ridges, rain-water pipes, and other external metallic parts of a building, may form part of the conductor employed for its defence; and it is a reprehensible instance of wanton expenditure, when conducting rods of thirty or forty feet long are placed by the side of a thick lead pipe of equal extent. In arrangements of this kind it is however necessary to observe, that the perpendicular conductor should pass in as direct a line as possible from the top of the house to the bottom.

The conductors for powder magazines are usually placed at a short distance from the building; when this method is adopted, the conducting rods should rise eight or ten feet above the highest part of the magazine, and penetrate as much below its foundation.

For the provision of perfect security from all possible danger, it has been proposed by Mr. Morgan to "attach to the sides of the foundation of each partition-wall, a strip of lead connected with a similar strip that entirely surrounds the foundation of the building. A perpendicular strip, on each side of the house, should rise from this bed of metallic conductors; and being connected with water-pipes, &c., be continued to the roof, where the method of guarding the bottom should be imitated. The top should be surrounded by a strip, whose connexion must spread over every edge and prominence, and hence continue to the summit of each separate chimney."*

The protection of the chimneys is of particular importance, for to these the discharge is most frequently determined. It has been said, that this circumstance arises from the conducting power of hot air; but this cannot be true, for no hot air issues from a chimney without a fire, yet these are frequently struck. The true cause is to be found in the superior conducting power of the charcoal, or soot, with which all chimneys that have been used are necessarily lined; for over a surface of this kind an electrical discharge will pass to a considerable distance.

The method of defence proposed by Mr. Morgan may be objected to as considerably expensive; for the strips of lead should be two inches wide, and a fourth of an inch thick; but he has justly observed, that if a proper use be made of the leaden pipes and copings that belong to most houses, and the common plumber or blacksmith is employed to fix the conducting strips, (which requires no other care

* Morgan's Lectures, vol. ii. p. 297.

than that they be secure, and well connected with each other,) a considerable proportion of this expense may be avoided. Hence, on the whole, this method may not be objectionable, where, from peculiarity of situation, a building is much exposed.

The rods employed for conductors may be fastened to the wall by iron or copper staples, considerably larger than the rod; and the part of it that passes through them, should be covered with two or three thicknesses of woollen cloth dipped in melted pitch; this serves to insulate the rod more completely from the building.

Independent of the protection required for the mast of a ship, it would be expedient to surround the deck with a strip of metal, which should be connected with the copper bottom; or if the ship is not coppered, the strip should be continued over the bottom or side of the keel, and be connected with others embracing the sides of the ship. The conductor or conductors for the masts, are to be connected with these metallic strips; and it then appears imposible for injury to occur.

Carriages are usually filletted round with metal; if these fillets are connected with each other, a covered carriage appears to be a sufficiently secure situation.

As a provision for personal security during a thunder storm, a few precautions are necessary. In the open air, shelter should not be sought immediately under a tree or building; for should they be struck, such a situation is particularly dangerous. The distance of twenty or thirty feet from tall trees or houses is, on the contrary an eligible situation; for should a discharge take place, these prominent bodies are most likely to receive it, and the less elevated bodies in their neighbourhood may therefore escape uninjured.

It is quite essential, during a thunder storm, to avoid every considerable mass of water, and even the streamlets that may have resulted from a recent shower; for these are all excellent conductors, and the height of a human being, when connected with them, is very likely to determine the course of an electrical discharge.

The partial conductors, through which the lightning directs its course when it enters a building, are usually the appendages of the walls and partitions; the most secure situation is therefore the middle of the room, and this situation may be rendered still more secure by standing on a glass-legged stool, a hair mattress, or even a thick woollen hearth rug.—The part of every building least likely to receive injury is the middle story, as the lightning does not always pass from the clouds to the earth, but is occasionally discharged from the earth to the clouds; hence it is absurd to take refuge in a cellar, or in the lowest story of a house; and many instances are on record in which the basement story has been the only part of a building that has sustained severe injury, the electric charge being divided and weakened as it ascended. Whatever situation be chosen, any approach to the fire-place should be particularly avoided; for (as it has been already stated) the chimneys are very likely to determine the course of the lightning. The same caution is necessary with respect to gilt furniture, bell wires, and moderately extensive surfaces of metal of every description. In a carriage, the precaution of keeping at some inches distant from its sides or back is also adviseable.

I have been rather ample in the enumeration of these particulars, from a desire to supply what useful information the present state of our knowledge may afford, and to diminish that anxiety and fear by which the minds of many are oppressed during the occurrence of these magnificent but awful phenomena. The operation of mysterious agents has always considerable effect on the human mind; and, by conjuring up a host of unreal terrors, may suspend the action of its best energies: but when we are enabled to ascertain the nature and influence of the powers which surround us, we are prepared to meet their effects with feelings equally remote from vain confidence and useless fear; and are thus enabled to avoid their dangers effectually and without agitation.

CHAP. II.

On the Phenomena of Thunder Storms, and on the probable Sources of Atmospherical Electricity.

THE experiments that have been detailed demonstrate, incontrovertibly, that electricity is identical with the phenomena of thunder and lightning; but its precise operation in the production of these impressive effects is by no means clearly understood. The circumstances most easily explained are the concomitants of this natural display of electrical power; for they do not differ materially in character from the phenomena of the electrical apparatus, and the extensive scale on which they are produced, as well as the circumstances of their occurrence, are ample sources of more considerable variety.

The spark and its attendant report, observed when a jar or battery is discharged, are effects perfectly analogous to a flash of lightning and its consequent thunder; and the variety of sound which characterizes this last, is equally the characteristic of every single explosion when it reaches the ear from a distance. A piece of artillery, discharged in any situation where the surrounding objects present irregular sources of reverberation, produces an effect which might be mistaken for the roar of thunder. This I have observed in several open situations of irregular surface, particularly on Hampstead Heath; and it may also be noticed when the guns are fired in St. James's Park, most remarkably when the observer is situated between them and the buildings towards Whitehall.

An observation of the varieties of the natural phenomenon will confirm this idea. At sea, where there is no diversity in the reverberating objects, the sound is regular, and decreases in intensity at every reverberation, until it gradually dies away; but in other situations, where the bodies capable of reverberating are numerous and irregularly placed, a succession of sounds are heard, varying in loudness and duration with the situation, distance, and nature of the sources of reverberation, and having no relation to the interval of time. When the flash is immediately followed by the report, a single explosion of a peculiar crashing sound is usually all that is heard, for the discharge has then taken place very near the observer, and damage will in such cases most frequently be found to have occurred in the immediate neighbourhood. When the sound does not immediately follow the flash, the rumbling and irregular noise most frequently occurs, for the distance is then sufficient to render the reverberation (from its extent) the most prominent feature of the phenomenon.

The interval between the flash and the explosion offers data for calculating the distance of a thunder stroke. For light moves with such velocity, that the time it requires to traverse any ordinary distance may be considered as nothing; but sound travels only at the rate of 1142 feet in a second. The flash and the report are really contemporaneous, but the one is immediately seen, and the other requires a second of time to traverse 1142 feet. Consequently the interval that elapses between the flash and the report being multiplied by 1142, or this last by the number of seconds in the interval, will give the distance of the explosion from the observer in feet. Thus, suppose the flash of lightning to take place five seconds before the thunder is heard, then $5 \times 1142 = 5710$, or 1 mile 430 feet, which is the distance of the explosion from the observer. This distance, it may however be proper to state, cannot be considered as a measure of his removal from danger; for it is the measure of an explosion which *has* taken place, and those that follow may occur in very different situations; for thunder clouds are sometimes continued over a considerable extent of country, and two or more flashes are not unfrequently seen in distant parts of the atmosphere at nearly the same time.

When the spark which causes lightning is seen, it is of the zigzag form, assumed by all powerful sparks when they traverse a considerable space of air, and in this its natural exhibition, the spark sometimes traverses a most prodigious interval. The appearance of two distinct streams at no very considerable distance from each other is sometimes produced, when part of a long zigzag is concealed by an intervening cloud; and the sudden and universal flash, called sheet-lightning, results from the reflection of an explosion which is more completely concealed. Of this last kind also appear those bright flashes which occur on summer evenings, and are not accompanied by thunder; a circumstance which it is difficult to account for, unless it may arise from their great distance from the earth's surface.

By far the greater number of flashes of lightning are harmless discharges from one cloud to another, and the instances in which it strikes the earth are comparatively rare; hence it appears that the clouds, or different portions of the atmosphere, are oppositely electrified; and it has been conceived with good reason by Mr. Morgan, that when the lightning strikes the earth, the latter may merely act as a discharging rod to shorten the striking interval between two charged clouds. Mr. Morgan has indeed supposed, that the earth cannot have its natural quantity of electric fluid either increased or diminished, because it is a conductor; but it is surrounded by air, and consequently is an insulated conductor; and our experiments teach us, that insulated conductors may be rendered either positive or negative; therefore the same may be inferred relative to the earth.

Others have supposed, that when such discharges occur, they result from the electrified cloud, producing by its proximity the contrary electricity in the earth, but when the size of the largest thunder cloud is compared with our globe, it will be evident that such an opposition could produce no more effect than would result from the approximation of an excited stick of sealing-wax to an insulated mountain. Besides, we have experimental evidence that, during the occurrence of such phenomena, different parts of the atmosphere are in opposite states of electricity at the same time; and as these states are dependant on each other, the discharge cannot take place but by passing from the positive to the negative, either directly, or by the intervention of part of the earth, between them.

The different electrical state of different parts of the atmosphere, obtains principally in the masses of vapour or clouds that float in it; and the origin of this electricity, as well as the cause of its various changes, is probably to be traced to the mutability of these masses; for it has been seen, that change of form, heating and cooling, friction, and the contact of dissimilar bodies, are the artificial sources of electrical excitation; and the clouds experience in succession the operation of all these causes. That electrical changes are connected with the state of vapour in the atmosphere, is evident from a consideration of all the phenomena, as will be apparent from the following facts.

1st. The electrical phenomena of the atmosphere take place in all climates to the greatest extent, during or near to the period of greatest heat, when the operation of the sun's rays has occasioned a considerable accumulation of vapour.

2d. Where this cause operates to the greatest extent, as in the countries within the tropics, these natural electrical phenomena are produced on a scale of the most tremendous magnitude.

3d. When the natural source of evaporation is assisted by collateral causes, electrical changes occur with astonishing activity. The eruption of a volcano is almost constantly attended by vivid lightnings; and the regions that surround the extensive sands of Africa, where the action of the sun's rays is assisted by reflection from an arid soil, are remarkable for violent storms and tempests; the air roasted by its passage over these sands, producing a rapid evaporation of the first moisture it meets, and becoming thereby so loaded as to evolve copious showers on any sudden diminution of temperature.

4th. By the action of winds, currents of air of different temperatures are often mixed; and such as have been heated and charged with moisture, are suddenly cooled; by this process water is precipitated, and electrical changes almost constantly occur; such are the hurricanes and terrific lightnings produced by the harmattan on the coast of Guinea, when it comes in contact with the cool air on the surface of the ocean; and such are also the electrical phenomena of all high ranges of mountains, for they occasion a condensation of the heated and moist winds that pass over their frozen summits; hence the magnificent lightnings of the Cordilleras, and the corruscations of the Alps.

5th. Electrical changes are in every situation most frequent when the causes of evaporation and condensation suddenly succeed each other. Those who have made regular observations on the electricity of the atmosphere, have always observed the greatest diversity when a rapid succession of rain and sunshine occurs; and such variable weather is also the most frequently attended by thunder storms. Even the diurnal changes of heat and cold, produce a perceptible effect on the atmospherical electricity;

for, according to the observations of Mr. Read, it is most obvious during the morning and evening dew, and weakest in their intervals.

Although the connexion of the circulation of water in the atmosphere with the production of its electrical phenomena is thus clearly pointed out, the immediate nature of this relation is by no means obvious. Volta discovered, that when water is rapidly converted into vapour, it leaves the vessel from which it has been evaporated, negative; and if the ascending vapour be received on an insulated piece of metal, it appears positive.* Hence he concluded, that water in expanding has its capacity for electricity increased, and consequently receives it from such bodies as are contiguous: admitting this, the condensation of vapour must necessarily be attended with positive signs of electricity, and the circulation of this subtle fluid in the atmosphere would be analogous to, and attendant on the circulation of water. But

* This experiment may be made with great facility by placing a few lighted coals in a crucible, on the cap of a gold-leaf electrometer, and projecting a few drops of water on them, whilst an insulated tin funnel is placed about a foot or 18 inches above. The electrometer will be electrified negatively, and the insulated funnel positively.

it has been observed by De Sausure and others, that the electrical effects of evaporation are not uniform, being directly opposite when different vessels are employed, and scarcely at all perceptible when the evaporation proceeds slowly, as it does in nature. To this it may be replied, that by the employment of different materials for the evaporating vessel, counteracting causes are probably introduced, by admitting the interference of chemical action, or the contact of dissimilar bodies; and such anomalous results have not been observed to occur when the water is evaporated from any substance analogous to the usual materials of the earth's surface. As to the different intensity of the electrical signs when the evaporation is more or less rapid, it is rather favourable than inimical to the opinior for such, it has already been observed, is also the case in nature; and if the usual process of natural evaporation produced an effect for each drop of vapourized water, equally powerful with that obtained in the usual mode of conducting these experiments, the assigned source of atmospherical electricity would be by far too prolific, and perpetual thunder storms an almost unavoidable consequence. For it has been calculated, that about 5280 millions of tuns of water are probably evaporated from the surface of the Mediterranean in one summer's day.* And a more recent estimate considers the mean evaporation from the whole earth, as equal to a column of 35 inches, from every inch of its surface in a year; which gives 94.450 cubic miles of water, as the quantity that circulates through the atmosphere annually.†

The extent of these phenomena is therefore fully adequate to the production of all the observed effects of atmospherical electricity; and the constant proportion of the effects to the operation of their assigned cause, renders that idea extremely plausible: but the observations of M. De Luc ‡ militate rather strongly against it; they are not indeed opposed to the preceding statement, for that is an enumeration of facts; but they tend to overthrow the principal hypotheses that have been advanced to account for the production of atmospherical electricity.

* Cavallo's Natural Philosophy, vol. ii. p. 409.

† Thomson's Chemistry, vol. iv. p. 78; and Manchester Memoirs, vol. v. p. 360.

‡ Idés sur la Météorologie, tom. ii. p. 158; or Nicholson's Journal, vol. xxvii. p. 241, &c.

The attention this excellent philosopher has paid to every subject connected with Meteorology, the acuteness of his views, the accuracy and extent of his experiments, and the unremitted continuation of his inquiries, confer a value on his observations that it is impossible to appreciate too highly; and whilst it affords me considerable pleasure to acknowledge the general accuracy of the objections he has urged, and the insufficiency of the tenets to which they are opposed, it is with no small diffidence I decline an assent to some of the explanations he has proposed in lieu of them. The limits of this work will admit no extensive theoretical discussion, I must therefore refer the reader to the Idées sur la Meteorologie, and to M. De Luc's various papers in the Philosophical Transactions, and in Nicholson's Journal, for an acquaintance with his luminous views; which embrace the most important phenomena in Meteorology and the sciences connected with it.

The labours of M. de Luc, last referred to, sufficiently prove that we are at present quite unacquainted with the nature of the process by which the apparent circulation of water through the atmosphere is effected. Evaporation will go on for months, and the air appear still dry; and driest in its upper strata, where the ascending vapours are supposed to pass: and in a stratum of this kind, where there is no evidence of the existence of any adequate quantity of moisture, clouds will suddenly form, and produce violent rain, accompanied by thunder and lightning, frequently of long duration. Nor are these clouds produced by any sudden condensation, for the heat of the clouds themselves is sometimes greater than that of the air by which they are surrounded;* and clouds that have been formed in the day frequently disappear at night, when, from the increased condensation, their continuance or increase might have been expected.

None of the principles, then, that have been hitherto advanced are adequate to account for the formation of clouds and rain; we have no evidence that they result from the saturation of the atmosphere with moisture, for they are

* De Luc sur la Météorologie, vol. ii. p. 100. This observation appears to me anomalous, for all bodies that are condensed have their temperature raised; and, consequently, if clouds *are formed* by the *condensation* of vapour, their heat *should be* greater than that of the air by which they are surrounded.

frequently formed in apparently dry air; we cannot shew that they are produced by cold, for the most remarkable changes of temperature are sometimes unattended by such phenomena; nor can we conclude with Dr. Hutton that they are occasioned by the mixture of air of different temperatures, for they frequently occur in a part of the atmosphere unagitated by winds either above or below; and it is not easy to conceive a mixture of this kind that will account for the quantity of rain that is frequently discharged on a sudden from a calm atmosphere.

In what state then does vapour exist in the atmosphere, when it is thus shrouded as it were from our observation? and by what process is it enabled so suddenly to display its energies, and leave its concealment with the accompaniment of such tremendous phenomena?

To this question we can reply only by suppositions, which are probably as remote from the truth as those which have been exploded by the discovery of these facts.

Mr. De Luc thinks its essential to conclude, that water may be converted into atmospheric air, and that the production of clouds and vapours arises from the decomposition of this air; and he also conceives that the electric fluid is a composite body, consisting of light, heat, and a peculiar species of matter which are combined together during some chemical changes in the atmosphere, and thus give rise to its electrical phenomena.

To me it appears quite unnecessary to suppose, that when any power or substance is latent, it has necessarily changed its form; and no such supposition has been yet admitted in various decisive instances of the kind: heat, for example, when it passes from an appreciable to a latent state, is not conceived to have undergone any change in its identity, but to have passed into a state of combination, or natural distribution; and the same may be said of electricity, which, as it has been already shewn, is always latent when its natural attraction is balanced. Even the power of attraction itself may be considered in the same way; that power is evident when a stone *is* falling; it is latent when the stone *has* fallen; but it still exists, unchanged, though unobserved.

The density of the air decreases in proportion to its distance from the earth; its particles are consequently at the greatest distance from each other in the most elevated stratum; and this may possibly be one cause in the production of these intricate phenomena. The rays of the sun reach the upper part of the atmosphere first, yet there the effect they produce is most trivial. The earth is most probably the original source of atmospherical electricity; but its effects are most apparent in the higher parts of the atmosphere; may not therefore the greater or less distance of the particles of air have an influence on the action of the electricity, heat, and vapour, that exist between them? and is it not equally plausible, that the phenomena in question may be produced by the approximation or recession of these particles, as that they result from unknown combinations and decompositions?

Speculations of this kind are only useful as a stimulus to inquiry, and should therefore be always regarded with caution, and offered with diffidence: they are indeed more favourable to the progress of true knowledge, when proposed as questions for experiment to resolve, than when expanded into hypotheses for experiment to confirm. For it is an impolitic excitement of false confidence, to erect a massive superstructure on a basis of doubtful stability.

Although the immediate causes by which the various phenomena of the atmosphere are produced, are still far beyond our comprehension; yet the connexion of their several effects is a sufficient demonstration that they are not purely mechanical, but subservient to the direction of supreme power and intelligence. By this means the most simple arrangement becomes the source of sublime effects. The process of evaporation which modifies the action of the sun's rays, and conveys to every part of the earth's surface a source of fertility, at the same time diversifies the appearance of the atmosphere by an endless variety of imagery, enlivens the horizon with the most brilliant and glowing tints, and probably effects those electrical changes, which are the precursors of the most magnificent phenomena in nature.

CHAP. III.

On some luminous Phenomena of the Atmosphere, the Observation of Atmospherical Electricity, and the Arrangement of a new System of Insulation.

THE satisfactory demonstration of the agency of electricity in the production of thunder and lightning, has occasioned the application of its principles to the solution of other phenomena, and in some instances it applies more plausibly than any other existing explanation. Such is particularly the case with some of those luminous appearances that occasionally enliven our atmosphere; effects, which have always been, and still continue amongst its most mysterious phenomena.

The Aurora Borealis, or northern light, is a phenomenon of this kind, whose appearance so exactly resembles some of the effects of artificial electricity, that those who have had the opportunity of comparing it with them, can entertain no doubt that their causes are identical.

When electricity passes through rarified air, it exhibits a diffused luminous stream, which has all the characteristic appearances of the northern lights. There is the same variety of colour, and intensity; the same undulating motion, and occasional corruscations; the streams exhibit the same diversity of character, at one moment minutely divided in ramifications, and at another beaming forth in one body of light, or passing in distinct broad flashes; and when the rarefaction is considerable, various parts of the stream assume that peculiar glowing colour which occasionally appears in the atmosphere, and is regarded by the uninformed observer with astonishment and fear.

There is therefore little doubt, that the natural phenomenon is occasioned by the passage of electricity through the upper regions of the atmosphere. The lowest estimate that has been made of the distance from the earth's surface at which it occurs, is that of Mr: Cavendish, who states that distance to be 71 miles; now, at 70 miles, the air is 1048576 times more rare than at the surface of the earth; and this is a degree of rarefaction beyond the power of any air-pump yet constructed.

These circumstances tend almost to a demonstration, that the light of the aurora is produced by the same means as the light of electricity; but there are other characteristics of that remarkable phenomenon that still remain unexplained.

Dr. Halley has described very fully the appearance of a remarkable Aurora, and has collected together a variety of observations that may serve as a history of its phenomena;* he ascribed its production to the same influence as that which causes magnetism; and the observations of Mr. Dalton prove that the direction of the luminous beams of the aurora is really that of the dipping needle.† Signior Beccaria conceives that the phenomena of magnetism are occasioned by a constant natural circulation of the electric fluid from north to south, originating from several sources in the northern hemisphere. The aberration of the common center of these currents from the north point, he supposes, may cause the variation of the needle; the period of this declination from the centre

* Phil. Trans. vol. xxx, p. 1099, or Motte's Abridgment, vol. ii. p. 116.

† Dalton's Meteorological Essays, p. 157.

may be the period of the variation; and the obliquity of the currents the cause of the dip of the magnetic needle.

The northern lights are at present very rarely visible here; a very few years since I observed them several times, and on one occasion their appearance was remarkably brilliant, and very similar to that noticed by Mr. Dalton in a phenomenon of this kind, which appeared on the 13th of October, 1792. An extract from his description will convey a general idea of an effect we have but rarely an opportunity to observe.

"Attention was first excited by a remarkably red appearance of the clouds to the south, which afforded sufficient light to read by, at eight o'clock in the evening, though there was no moon, nor light in the north. Some remarkable appearance being expected, a theodolite was placed to observe its altitude, bearing, &c.

"From 9½ to 10 P. M. there was a large, luminous, horizontal arch to the southward, almost exactly like those we see in the north; and there was one or more faint concentric arches northward. It was particularly noticed, that all the arches seemed exactly bisected by the plane of the magnetic meridian. At half past 10 o clock, streamers appeared very low in the S. E. running to and fro, from W. to E., they increased in number, and began to approach the zenith, apparently with an accelerated velocity; when, all on a sudden the whole hemisphere was covered with them, and exhibited such an appearance as surpasses all description. The intensity of the light, the prodigious number and volatility of the beams, the grand intermixture of all the prismatic colours in their utmost splendour, variegating the glowing canopy with the most luxuriant and enchanting scenery, afforded an awful, but at the same time the most pleasing and sublime spectacle in nature. Every one gazed with astonishment; but the uncommon grandeur of the scene only lasted about one minute; the variety of colours disappeared, and the beams lost their lateral motion, and were converted, as usual, into the flashing radiations; but even then it surpassed all other appearances of the *aurora*, in that the *whole* hemisphere was covered with it.

"Notwithstanding the suddenness of the effulgence at the breaking out of the *aurora*, there was a remarkable regularity in the manner.—

Apparently a ball of fire ran along from east to west, and the contrary, with a velocity so great as to be barely distinguishable from one continued train, which kindled up the several rows of beams one after another: these rows were situate before each other with the exactest order, so that the bases of each row formed a circle crossing the magnetic meridian at right angles; and the several circles rose one above another in such sort, that those near the zenith appeared more distant from each other than those near the horizon, a certain indication that the real distances of the rows were either nearly or exactly the same. And it was further observable, that during the rapid lateral motion of the beams, their direction in every two nearest rows was alternate, so that whilst the motion in one row was from E. to W., that in the next was from W. to E.

The point to which all the beams and flashes of light uniformly tended, was in the magnetic meridian, and as near as could be determined, between 15 and 20 degrees south of the zenith. The *aurora* continued, though diminishing in splendour, for several hours. There were several meteors (falling stars) seen at the time;

they seemed below the aurora, and unconnected therewith."*

When the northern lights are visible in this country, they are said to appear chiefly in the spring and autumn, and usually after a period of dry weather; they do not refract the light of the stars, which are often distinctly seen through the luminous arch or beams.

They are seen more rarely in countries nearer the Equator, but occur almost constantly during the long winters, in the polar regions, and with a lustre of which we can form but a faint conception.

In the Shetland iles they are called "merry dancers," and are the regular attendants of clear evenings, giving a diversity and cheerfulness to the long winter nights. Their first appearance is at twilight, just above the horizon; they have at first no particular brilliance or motion, but after some time break forth into streams of refulgent light, whose Protean columns gradually assume every possible variety of form, and shade of colour; frequently covering the whole visible hemisphere; which then presents the most brilliant spectacle imagination can conceive.

* Dalton's Meteorological Essays, p. 65.

In Hudson's Bay, the refulgence of the aurora is stated to be frequently equal to that of the full moon. In the northern latitudes of Lapland and of Sweden their brilliance is so remarkable and constant, as to enliven the path of the traveller during the whole night. In the north-eastern parts of Siberia they are also described as moving with incredible velocity, and clothing the sky with a most brilliant luminous appearance, " resembling a vast expanded tent, glittering with gold, rubies, and sapphire." This phenomenon is said to be accompanied by a loud hissing or crackling noise, so terrific, that when the fox hunters, on the confines of the icy sea, are overtaken by it, their dogs lie close to the ground, and refuse to move until the noise has passed.

That a noise of this kind occasionally accompanies the northern lights, has been testified by several observers, and amongst others by Mr. Nairne and Mr. Cavallo; the last says, he has heard it distinctly on several occasions. This effect is the most extraordinary of all that accompany these phenomena, and if established as a fact, is perfectly unaccountable; for, from the extent of country over which the aurora is frequently seen, it is certain that it must occur at a very considerable height above the earth's surface; and though the calculations on this subject differ remarkably,* yet the very lowest estimate gives an elevation, at which, according to the known principles of philosophy, there exists no medium capable of transmitting sound!

An aurora has been sometimes observed near the south pole, which seems to favour Beccaria's idea, that its cause is the circulation of a fluid. Its appearance is similar to that of the northern light, but without the same diversity of colour.

The beams of the aurora appear to converge towards the zenith, and their summits seem narrower than their bases; but this, as was observed by Dr. Halley and Mr. Cavendish, is merely an optical illusion; and it has been shewn by Mr. Dalton, that the beams are really cylindrical, and parallel to each other; and that

* The height of an aurora was determined by Boscovich at 825 miles. By Bergman, from a mean of thirty computations, at 468 miles. By Euler, its height is estimated at several thousands of miles. By Mairan, at 200 leagues. Mr. Cavendish, by a comparison of observations, states it at from 50 to 71 miles. Mr. Dalton and Mr. Crosthwaite, at 150 miles.

the distance of their bases from the earth, is equal to, or probably greater than the length of the beams: and he has calculated that these beams are 75 miles long, and $7\frac{1}{2}$ miles diameter.*

The other luminous appearances of the atmosphere which have been attributed to electricity, are those usually called Meteors, or Fire-Balls. Of these there are two very distinct classes; the one of considerable apparent size, moving *progressively* over a very considerable space, and sometimes dispersing in divided sparks.† These occur but rarely, and their dispersion is sometimes attended by the fall of stones; a fact which renders the application of electrical principles quite insufficient to explain them. Until lately, the statements of the fall of meteoric stones were but slightly credited; but considerable attention having recently been bestowed on the investigation of these accounts, and the examination of the stones said

* Meteorological Essays, p. 177.

† A remarkable meteor of this kind occurred in August 1783. It was observed by Mr. Cavallo, from the terrace of Windsor Castle, and is described by him in the Philosophical Transactions for 1784, article 9th; and also in the 4th volume of his Natural Philosophy, p. 359.

to have fallen; the fact appears to be established, and with the remarkable circumstance, that all the stones of this kind hitherto examined are of nearly similar composition.

In the present state of our knowledge no satisfactory explanation of these meteors can be given: all that has been proposed on the subject is perfectly visionary; and the details of vague speculation would be but tedious and useless. Electricity is inadequate according to our present acquaintance with its agencies, and some of the concomitant phenomena indicate, almost to a demonstration, that combustion is concerned in the production of these appearances.

The second kind of meteors occur much more frequently; they are usually called falling stars, or shooting stars; they vary somewhat in size and colour, but have nearly the same rapidity of motion, moving swiftly in various directions, but chiefly inclining towards the earth. They occur in various states of the atmosphere, but most frequently when the observed causes of electrical change have been most active; hence they are found to prevail in clear frosty nights, and at other times when there are dry easterly winds, and a clear sky;

they also abound in the clear intervals of showery weather, and on those summer evenings when well-defined clouds are floating in a clear atmosphere. I have observed them most frequently at times when considerable extraneous light prevailed, either from the clear refulgence of the stars and planets, or from the presence of the moon; but I have also noticed them on partially cloudy, and dark nights.

Various are the observations that prove their occurrence during the splendid display of the northern lights; and it is a remarkable fact, that they always appear lower than those lights, which seems to indicate that they are in all probability the same matter moving through a more resisting medium.

In favourable states of the atmosphere these appearances succeed each other with such rapidity, that I have frequently counted 30 in the space of a single hour; and on some occasions nearly twice that number. The frequency of their occurrence indicates, that they are produced by some of the usual atmospherical changes; and the circumstance of no meteoric stones being produced by them, renders it almost certain, that their nature is materially different from that of the large meteors.

Independent of other sources of analogy, the following are striking arguments in favour of their electrical origin.

1st. The light of falling stars is similar to the light of the electric spark.

2d. They occur as frequently and as irregularly as other electrical changes in the atmosphere.

3d. Their motion, like that of electricity, is inconceivably rapid; and the longest interval they strike through, is traversed in an interval of time too minute to admit of measurement.

4th. They occur most frequently, during, or near to those changes of weather, that are known to influence the electrical state of the atmosphere.

5th. Their direction is never constant; they occur in every part of the atmosphere, and move in almost every variety of inclination; such is also the case with lightning.

6th. The appearance of falling stars may be accurately imitated by electricity; and the circumstances on which the success of such experiments depend, are such as are likely to occur in the production of the natural phenomenon.

If electricity be passed through an exhausted receiver gradually, it assumes the appearance of the northern lights; but if a considerable electrical accumulation be suddenly transmitted, it will pass through the receiver with all the straightness and brilliance of a falling star. If the receiver is six inches diameter, and fourteen or sixteen inches high, the full charge of a moderate sized jar is necessary to produce this effect, and it occurs most readily when the receiver is but moderately exhausted, so that the rarefied air it contains may have some degree of resistance.

The artificial imitation of the two phenomena, therefore, require the same conditions for their production as appear to obtain in nature; for the aurora occurs in the highest parts of the atmosphere where the air is most rarefied; and the most accurate imitation of its appearance, is obtained in the most perfectly exhausted receiver; falling stars take place much lower, where the air has more density, and to imitate them, it is necessary to employ a medium that opposes some resistance.

These facts are confirmed by almost every possible variation of the experiment, and in some instances the approach to the appearance of the natural phenomenon is remarkably striking; for the electric fluid may be made to pass over a very considerable interval by the employment of a proper apparatus.

I employ for this purpose a glass tube, five feet in length, and 5-8ths of an inch diameter, capped with brass at each extremity. When this tube is exhausted, no ordinary spark will pass through it in any other than a diffused state, but by employing the charge of a very large jar a brilliant spark is obtained through the whole length of the tube. Mr. Morgan found, that in a shorter tube of the preceding description, the appearance of a falling star was produced by a spark which would pass through ten inches in the open air, provided the tube did not exceed forty-eight inches in length, and contained a quantity of air, which under common circumstances would have filled one-twenty-fourth of its capacity; but if this small quantity of air was further dilated by the action of an air-pump, the most powerful spark would pass through it in a divided stream. By employing a very narrow tube of the same length, the confined column of rarefied air resisted the charge sufficiently to produce the appearance of the brilliant spark through its whole length, whenever the accumulated electricity was sufficiently powerful to pass through it.

These experiments, and the analogies by which they are supported, render it highly probable that electricity is connected with the appearance of these lesser meteors; but neither the precise mode of their production, nor the purpose they answer, is as yet by any means explained.

In a very electrical state of the atmosphere a luminous appearance is sometimes observed on the summits of spires and the masts of ships, and a similar effect has been occasionally noticed on the points of spears; it is analogous to the light that appears on any slender and prominent conductor, when it is surrounded by electrified air, or approximated to an excited electric.

Earthquakes, water-spouts, and even volcanoes have been ascribed to the agency of electricity; but the idea is not supported by any other circumstance than the occasional occurrence of some electrical effects during the operation of such phenomena; and it is much more probable that these are a consequence of their action, than an evidence of the cause by which they are produced.

The electrical phenomena of the atmosphere are not confined to its luminous effects, for it has been found that the air is almost constantly electrified; and observations have been made on the character and mutations of its phenomena, by the aid of kites, insulated rods, and extended wires. It may be interesting to give some account of the structure and arrangement of these several sources of inquiry.

An electrical kite should be constructed in the most simple manner, for it is an apparatus very liable to be injured or lost; its size should be moderate, as there is not often sufficient wind to raise one that is very large, which is besides, on several other accounts, very troublesome to manage. An ordinary paper kite, about four feet in height, and two feet wide, varnished with drying oil to defend it from the rain, is sufficiently well adapted for this purpose. The string must be made with a thin copper or silver thread (such as is used for gilt lace) entwisted with the twine of which it is formed through its whole length. When the kite is raised, the string is insulated by attaching to it a silk cord, whose opposite extremity may be fastened to a rail, or any fixed or heavy body. The end of the metallic string is to be connected with an insulated conductor, and at two inches from the extremity of this conductor a brass ball, well connected with the ground or the nearest water, is to be placed; so that when the electricity is sufficiently intense to pass an interval of two inches, it will be conducted safely away without injury to the experimentalist, who should be cautious, in such cases, not to approach the insulated conductor; but if he has occasion to remove any apparatus to or from it, to do so by the aid of long insulating handles or forceps. By an attention to these circumstances, M. de Romas was enabled to manage with security the very formidable accumulation already stated to have occurred in his experiments; but as no useful purpose appears likely to be accomplished by such temerity, I should rather advise, on every similar occasion, an immediate retreat to a secure distance from the apparatus.

In raising or lowering the kite, the shocks that are sometimes inadvertently received may be effectually prevented by suffering a part of the string between the operator and the kite to bear constantly against the brass ball that is connected with the ground; and this precaution is very essential when thunder clouds are over head; but on such occasions, it is quite unnecessary to raise the kite, since the atmospherical electricity may then be observed by more simple means.

The effects obtained by an electrical kite are usually greater in proportion to the length of the string; and when the atmospherical electricity is very weak, it has been sometimes found necessary to employ two or three kites, one above the other, that a sufficient length of conducting cord might be exposed to the air. These additional kites have each a long slit through the middle stick, or straiter; and when the first kite has taken as much cord as it will carry, the end is passed through the slit of a second kite and tied to its string; and when this has arisen as high as it will, a third kite may be added in the same manner; but it is to be observed, that the opposite currents of air that frequently prevail at different heights in the atmosphere are very apt to interfere with the success of an experiment of this kind.

It is evident that the kite serves merely to extend a length of conducting cord in the atmosphere, and as it is not suited for permanent observations, other means have been employed for that purpose. Signior Beccaria extended a long wire permanently between the top of a cherry-tree and the summit of a long pole attached to a stack of chimneys. Its extremities were insulated by glass covered with sealing-wax, and defended from the rain by small funnels of tin. A branch proceeded from this wire through a pane of glass into his room, where observations were made on the electricity collected by this apparatus, and its indications compared with the action of the hygrometer, and with other concurrent phenomena.

The wire usually extended was 132 French feet in length; it was placed on the top of the Hill of Garzegna, in the vicinity of Mondovi; an elevated situation, from which the whole compass of the Alps and the plain of Piedmont is perceptible. At one time he also stretched an insulated rope, of 1500 Paris feet long, over the river Po, which exhibited intense signs of electricity whenever a shower was falling.

The observations of this assiduous philosopher were continued above fifteen years; they prove that the atmosphere is almost constantly electrified, and that its electricity is usually positive, and has a manifest relation to the state of the vapour it contains: the electrical indications of the apparatus are frequently affected by the passage of clouds over it, and by the transition of a current of air from any situation where clouds are forming or vapours falling. With the exception of the action of circumstances of this kind, negative electricity is rarely observed in the atmosphere, and it appears therefore probable, that when it does occur, it may result from the action of the strong positive charge of one part of the atmosphere on the natural electricity of another contiguous portion; or, in other words, that the signs then produced result from the influence of the permanent atmospherical electricity, and not from its actual communication.

My friend, Andrew Crosse, Esq. of Broomfield, near Taunton, a most active and intelligent electrician, has lately made very numerous observations with a remarkably extensive atmospherical conductor, consisting of copper wire one-sixteenth of an inch thick, stretched and insulated between stout upright masts of from 100 to 110 feet in height. The most unwearied exertion has been employed to give unexampled extent and perfection to this apparatus; the insulated wire has been extended to the extraordinary length of one mile and a quarter; and a variety of ingenious contrivances have been applied to preserve the insulation; but the length of the wire rendered it so liable to injury, and subject to depredation, that it has been found expedient to shorten it to 1800 feet; and until the present time no means have been devised that sufficiently preserve the insulation during a dense fog or driving snow.

There are some minor inconveniences attendant on the use of this apparatus, which are obviated by fixing it very securely, and providing a contrivance by which it can readily be raised or lowered to cleanse the insulators; for these are sometimes rendered conducting by spider's webs; and the secure fixing of the wire is essential to resist the weight of innumerable swallows that occasionally perch upon it, and of wood pigeons and owls, which frequently fly with considerable force against it.

A wire of this kind has been kept strained for eighteen months without injury; and from the observation of its indications, and those obtained in other experiments of less duration, the following deductions have been made.

1st. In the usual state of the atmosphere, its electricity is invariably positive.

2d. Fogs, rain, snow, hail, and sleet produce alterations of the electric state of the wire: it is usually negative when they first appear, but oftentimes changes to positive, increasing gradually in strength, and then gradually decreasing and changing its quality every three or four minutes. These phenomena are so constant, that whenever the negative electricity is observed in the apparatus, it is considered as certain there is either rain, snow, hail, or a mist in its immediate neighbourhood, or that a thunder-cloud is near.

3d. The approach of a charged cloud produces sometimes positive and at others negative signs at first; but, whatever be the original character, the effect gradually increases to a certain extent, then decreases, and disappears, and is followed by the appearance of the opposite signs, which gradually extend beyond the former maximum, then decrease, terminate, and are again followed by the original electricity. These alternations are sometimes numerous, and are more or less rapid on different occasions; they usually increase in intensity at each repetition, and at last a full dense stream of sparks issues from the atmospherical conductor to the receiving ball, stopping at intervals, but returning with redoubled force. In this state a strong current of air proceeds from the wire and its connected apparatus; and none but a spectator can conceive the awful though sublime effect of such phenomena. At every flash of lightning an explosive stream, accompanied by a peculiar noise, passes between the balls of the apparatus, and enlightens most brilliantly every surrounding object, whilst these effects are heightened by the successive peals of thunder, and by the consciousness of so near an approach to its cause.

During this display of electric power, so awful to an ordinary observer, the electrician sits quietly in front of the apparatus, conducts the lightning in any required direction, and employs it to fuse wires, decompose fluids, or fire inflammable substances; and when the effects are too powerful to attend to such experiments securely, he connects the insulated wire with the ground, and transmits the accumulated electricity with silence and with safety.

4th. A driving fog, or smart rain, frequently electrifies the apparatus nearly to the same extent as a thunder cloud, and with similar changes.

5th. In cloudy weather weak positive electricity usually prevails: if rain falls it frequently changes to negative; but the positive state is resumed when the rain ceases.

6th. In clear frosty weather the positive electricity is stronger than in a fine summer's day. The intensity of the electrical signs at different seasons is expressed, in descending order, in the following list, commencing with that whose effects are most considerable.

1. During the occurrence of regular thunder-clouds.
2. A driving fog, accompanied by small rain.

3. A fall of snow, or a brisk hail-storm.
4. A smart shower, especially on a hot day.
5. Hot weather succeeding a series of wet days,
6. Wet weather following a series of dry days,
7. Clear frosty weather, either night or day,
8. Clear warm summer weather.
9. A sky obscured by clouds.
10. A mackarel-back, or mottled sky.
11. Sultry weather, the sky covered with light hazy clouds.
12. A cold damp night.

To this may be added, as least electrical of all, a peculiar state of the atmosphere which sometimes occurs during the prevalence of north-easterly winds; it is characterized as particularly unhealthy, and is remarkable in producing a sensation of dryness and extreme cold, which is not accompanied by a correspondent depression of the thermometer.

The usual positive electricity is weakest during the night; it increases with the sun rise, decreases toward the middle of the day, and increasing as the sun declines, it then again diminishes, and remains weak through the night. This fact is one of the most instructive resulting from these observations, and is confirmed by most of the regular experiments on atmospherical electricity that have been made; it clearly proves that the electricity of the atmosphere is influenced by the same causes that promote the equal distribution of moisture.

A very regular series of observations on the electricity of the atmosphere, have been also made by Mr. Read of Knightsbridge. His apparatus consisted of a deal rod 20 feet long, which was secured very firmly at the bottom by supports of glass covered with sealing wax, within a room in the upper story of his house; the insulation was by this means better preserved than when freely exposed in the open air. The upper extremity of the rod passed through the centre of a hollow wooden cylinder fixed through the cieling and the roof; and the interior of the cylinder was defended from the rain by a large tin funnel affixed to the rod at a little height above its upper extremity.* The observations made during two years with this

* A description of the apparatus, and a journal of the observations made with it, may be seen in the 81st volume of the Phil. Trans. p. 185, &c. or in Read's Summary View of Spontaneous Electricity, p. 103.

apparatus, agree very nearly with the preceding deductions.

For temporary or occasional observations, very simple contrivances may be employed. A common jointed fishing-rod having a glass stick covered with sealing-wax substituted for its smallest joint, may be occasionally projected out of the upper window of a house. A pair of pith balls must be attached to a cork in which the end of the glass stick is thrust; and this part of the apparatus is to be occasionally uninsulated, by placing a pin in the cork, connected with a thin wire held in the hand. In this uninsulated state, the fishing-rod and its attached electrometer are to be held for a few seconds projecting from the window, and whilst in this position the pin is to be withdrawn by pulling the thin wire; this insulates the electrometer, which may be then drawn in and examined. Its electricity will be contrary to that of the atmosphere.

Mr. Bennet recommends a tin funnel insulated by means of cement at the extremity of a long rod. A wire is to proceed from the funnel to a sensible electrometer placed within doors; and when the atmospherical electricity is weak, a torch or small lamp may be attached to the funnel, for flame facilitates the collection of electricity from air.

In every arrangement of this kind, the principal difficulty is the preservation of the insulation; and when minute differences from the electrical standard are to be investigated, this difficulty is productive of the most serious inconvenience. Insulation may, (as it has been before noticed,) be partially preserved by coating all the glass insulators with sealing-wax; but this supplies only a temporary defence, moisture is eventually precipitated on them, and in removing this, it is scarcely possible to avoid exciting the surface of the wax, which, by producing a new source of electricity, renders the result of every delicate experiment equivocal. I have been successful in an attempt to obviate this inconvenience to a very considerable extent, by a new arrangement of those parts of any delicate electrical apparatus on which the permanence of its insulation may depend.

Reflecting that the perfection of insulators is constantly diminished by the deposition of moisture from the atmosphere on their surfaces, and that this moisture exists therein diffused as one gas mixed with another; it seemed to follow, that if the contact of the atmosphere with the insulators was less free, their insulation would be longer preserved, as the transition of moisture from it to them would be necessarily retarded. It was obvious this might be effected by enclosing the insulator within a narrow channel, as the air in contact with it would be then limited in quantity and little disposed to motion; for all gases communicate slowly with each other when separated by narrow tubes, and slower in proportion as these are less in diameter, and of greater extent.

The application of this principle to the perfection of the gold leaf electrometer was the first trial of its excellence; and the result was the most satisfactory demonstration of its utility.

The instrument is constructed, as usual, with a glass cylinder surmounted by a wooden or metal cap. The insulation is made to depend on a glass tube of four inches long, and one-fourth of an inch internal diameter, covered both on the inside and outside with sealing-wax, and having a brass wire of a sixteenth or twelfth of an inch thick, and five inches long, pass through its axis so as to be perfectly free from contact with any part of the tube, in the middle of which it is fixed by a plug of silk, which keeps it concentric with the internal diameter of the tube. See fig. 33. A is a brass cap screwed upon the upper part of this wire; it serves to limit the atmosphere from free contact with the outside of the tube, and at the same time defends its inside from dust. To the lower part of the wire the gold leaves are fastened. The glass tube passes through the centre of the usual cap of the electrometer, and is cemented in it at about the middle of its length, as may be seen by the dotted lines which represent this cap. When this construction is considered, it will be evident that the insulation of the wire, and consequently of the gold leaves, will be preserved until the *inside* as well as the *outside* of the glass tube is coated with moisture; but so effectually does the arrangement preclude this, that some of these electrometers that were constructed in 1810, and have never yet been warmed or wiped, have still apparently the same insulating power as at first. The instrument is represented complete by figure 1.

Had this simple arrangement been found insufficient, it was my intention to lengthen the narrow channel, and thus increase the security of the insulator; for this might easily be effected to almost any extent, by enclosing a series of open tubes one within another, and securing each in its place by a plug of silk, touching the adjoining tube in one point only. In this way, by multiplying the tubes the most perfect possible insulation may be obtained.

For the insulator of a vertical atmospherical apparatus, a stick of glass, ten inches long and one inch diameter, coated with sealing-wax, may be capped with brass at each extremity. Each cap is to be furnished with a screw to receive the lid of a cylindrical tin funnel. There are to be two such funnels, one screwed at each end of the insulating pillar; they may be about eight inches long, and one smaller than the other, in such proportion that the circumference of the stick of glass and the two funnels may form a series of concentric circles, distant from each other about a quarter of an inch. The apparatus is represented by Fig. 34, the funnels being delineated by dotted lines. It is evident that, in this apparatus, the vapour must first traverse the space between the outer and inner funnels, and then the interval between the inner funnel and the stick of glass, before the insulation can be destroyed; and this space may be lengthened to any extent by increasing the number of concentric funnels. This arrangement is very simple and durable, and though the limit of its insulation is the distance of the funnels, that is a quarter of an inch; this will be found sufficient for the most essential observations on the atmosphere, and the higher intensities may be obtained if desired by prolonging the insulator to some inches below the cap of the lower funnel, as shewn in the figure; or by making this lower and internal funnel of a glass tube covered with sealing-wax.

If an apparatus of this description be used to insulate the horizontal wire, the open end of the larger funnel should have a circular tin plate of nearly twice its diameter placed opposite to it at a short distance, to prevent the intrusion of driving rain or snow.* Or, what might perhaps prove more effectual, it may be placed within a sort of pigeon house, having a hole in its side for the wire to pass through.

* See Figure 35.

This method of insulation is applicable to almost all the varieties of an electrical apparatus: my present limits will not allow a detailed statement of its modifications, but the ingenious electrician will not find it difficult either to comprehend or employ them; and he who does this will not fail to acknowledge and appreciate their value.

An exploring wire for atmospherical electricity has been insulated nearly agreeable to this plan by a very assiduous and promising electrician, F. Ronalds, Esq. of Hammersmith. The apparatus was erected in a field near Highbury Terrace, Islington, and continued in constant activity for several months: the insulation was tolerably well preserved, but not uniformly so; this he attributes, in part, to the hasty and probably imperfect construction of the apparatus, and partly to the insufficiency of the most perfect insulators, when the stratum of air between the wire and the ground is so moist as to become a conductor of electricity. I communicated to this intelligent friend a plan of the Chevalier Landriani, to mark the diurnal mutations of atmospherical electricity, by an insulated wire, connected with the conductor, and carried round over a resinous surface by means of the index of a dial; powders being afterwards projected on the resinous plate would mark, by the figures they assumed, the intensity and quality of the electricity that had prevailed during the different hours of the day. Mr. Ronalds proposes to substitute for the resinous plate a series of electrometers, constructed on my principle of insulation: a soft wire proceeding from the atmospherical conductor being connected with the index of an insulated dial, is to be so arranged that it will touch in succession the cap of each electrometer. These being perfectly insulated, will retain the electricity communicated to them, and being adjusted for separate hours, or other divisions of time, the various electricity of each period of a day, or any longer interval may be ascertained by one observation. This method may be readily put in practice; it will be very useful to register the changes (if any) that occur during the night, and promises to afford a tolerably accurate indication at any required time of the state of atmospherical electricity during the absence of the observer.

CHAP. IV.

Connexion of Electricity with Medicine, and with Natural History.

THE numerous extraordinary properties of this surprizing agent, occasioned the application of its powers at a very early period to various organized bodies; and the results observed, or imagined, gave rise to a variety of fanciful opinions, which are now referred to but as monuments of credulity and imposture.

The Abbe Nollet is said to have made the first experiments of importance on this subject, and they do not appear to have been extended or repeated with greater accuracy. From his results it may be deduced; 1st. that the conversion of evaporable fluids into vapour is promoted by electrifying them; 2d. that the motion of any fluid through a capillary tube is accelerated by electricity, and that the acceleration is comparatively greater the smaller the capillary tube; 3d. that the motion of any fluid through a tube of moderate size is not sensibly affected by electricity, neither is the circulation of the blood promoted or retarded thereby;* 4th. the insensible perspiration of animals is increased during the time they are electrified, and the same may be observed of vegetables.

These remarkable facts, (and such they really appear,) render it highly probable that the electric fluid may be active in the production of many phenomena of nature, that exhibit not the slightest trace of its usual effects: the supposition that electricity is connected with animal and vegetable life, has been derived almost wholly from their evidence, and its application as a medical agent, if proposed on rational principles, was in all probability derived from the same cause.

* This is confirmed by the experiments of Mr. Cavallo, and the more recent and extensive trials of Dr. Van Marum and Mr. Cuthbertson, with the large machine at Harlem; but Mr. Partington, whose experience as a medical electrician is considerable, assured Mr. Cavallo, that in a diseased state of the body, an evident acceleration of the pulse is often observed to result from the application of electricity. Mr. Carpue states, that having opened a vein from which the blood did not readily flow, he electrified the patient, and the blood then streamed forth freely.

Soon after the experiments of Nollet, accounts were received from Venice and Bologna of some pretended miraculous effects of electricity in medicine. It was asserted, that by that power, odours, and the medical action of drugs, might be transmitted through glass vessels, and iron chains; and, that to produce the most remarkable cures, nothing more was required than to place some simple drug in the patient's hand, whilst he was electrified, or to enclose the medicine in a cylinder or phial, and transmit its efficacy with the electric fluid to any required distance. These assertions were made with the decisive language of experience, and were apparently supported by respectable testimony; they were investigated with great care by the Abbé Nollet, who was at the expense of a journey to Italy for that express purpose; they were also examined by several members of the Royal Society, and by Dr. Bianchini of Venice. The result of these inquiries decisively proved that the whole of these pretended miracles were artful fabrications, entirely unsupported by experiments, and invented for the deception of the credulous, and the burthen and disgrace of science.

Too many instances have since occurred of similar absurd pretensions to the cure of diseases by the influence of imaginary natural powers. An atmosphere charged with electricity has been proposed as the vehicle of strength and vigour; and the motion of a pointed piece of metal has been represented as a panacea! To dwell on such mummery would be an undeserved attention to that empiricism, which should be forgotten, or despised.

The scientific application of electricity to medicine, has made less progress than the success which has attended it might have been justly expected to produce. It appears from almost every trial of its power hitherto made, that under judicious management its application can do no harm, and that in many of the most distressing disorders it has frequently been of considerable service. These are powerful recommendations, and when it is added, that it is an external, and by no means painful remedy; and that it may be applied immediately to the affected limb without interfering with any other part, its advantages must appear to be considerable.

My own experience, although comparatively

limited, sufficiently warrants the preceding opinion; even in those cases which I have considered as unsuccessful, some relief has usually been obtained, and it is probable that more continued attention than I have had the opportunity to bestow, would have been followed by more perfect results.

The machine employed for medical purposes should have sufficient power to furnish a constant stream of strong sparks, for in many cases an application of that kind is essential. If it is a plate machine, the diameter of the plate should not be less than from eighteen inches to two feet; if it is a cylinder, the diameter may be from eight to fourteen inches.

The auxiliary apparatus is very simple; the most essential are, 1st. A jar fitted up with Lane's electrometer,* by which shocks may be given of any required force. 2d. A pair of directors, each consisting of a glass handle, surmounted by a brass cap with a wire of a few inches in length, having a ball screwed on its extremity; this ball may be occasionally unscrewed and a wooden point substituted for it. When shocks are passed by the aid of these di-

* Figure 22.

rectors, they are applied at the opposite extremities of the part through which the charge is to pass, and being respectively connected by conducting wires, the one with the outside of the jar, and the other with the receiving ball of Lane's electrometer previously placed at the required distance, the jar may be set to the machine, which is then put in motion until any required number of shocks has been given.

The insulated director is also employed to give sparks, being held by its glass handle, and its ball previously connected with the conductor by a flexible wire being brought near the patient, or rubbed lightly over a piece of flannel or woollen cloth laid on the affected part. When the eye or any delicate organ is electrified, the ball of the insulated director is unscrewed and the wooden point applied, at the distance of about half an inch from the part. The stream of electrified air which passes from the point under such circumstances produces rather a pleasant sensation.

Very excellent flexible conductors for medical purposes, may be made by sewing a thin spiral brass wire (such as is used for braces,) within a thick silk ribbon.

3d. An insulated stool, (that is, a stool with glass legs,) is sometimes employed; it should be of sufficient size to receive a chair upon it, with a resting place in front of the chair for the feet. The patient being placed on the insulated chair, and connected with the conductor of the machine, becomes a part of it, and sparks may be drawn from any part of the body by a person who stands on the ground and presents a brass ball to it. If the ball is held by a wooden handle, the sensation is less painful than when it is held by metal.

It has been doubted by Mr. Morgan whether any of the minor applications of electricity can be at all effectual, for he supposes that the electric fluid in such cases passes only over the surface of the skin, and not through the body. But this objection is purely hypothetical, for we know not in what manner conductors like the human body transmit slight accumulations of the electric fluid; nor have we any correct idea of the principle on which the medical powers of electricity depend; but experience has most decisively proved, that in its mildest form, (that of the current of electrified air from a point,) it has frequently effected very remarkable cures.

The most instructive collection of cases of the medical application of electricity I have yet seen, may be consulted in a very neat "Introduction to Electricity and Galvanism," published by Mr. Carpue. The unsuccessful cases are given, as well as those that were more fortunate; and from a personal knowledge of the science, talent, and impartiality of Mr. Carpue, I am satisfied that a better authority cannot be referred to.

A brief enumeration of some instances of disease, in which the application of electricity has been beneficial, it may be proper to insert.

1st. *Contractions.* Those only that depend on the affection of a nerve; and in many of these it has been employed without effect, whilst in others of long duration immediate relief has been obtained.

2d. *Rigidity.* Very frequently relieved, but usually requiring some perseverance in the application, to complete the cure.

3d. *Sprains, Relaxation, &c.* Electricity may be applied in all these cases with good effect, but its application should be deferred until the inflammation has subsided.

4th. *Indolent Tumors.* Strong sparks, and

slight shocks, are often effectual. The most numerous cases are those of scirrhous testicle; and there are some instances of the successful dispersion of scirrhous induration of the breast. Ganglions have also been removed from the wrists or feet by the frequent application of sparks.

5th. Mr. Carpue states, that electricity is a good preventative against chilbains; and mentions two instances in which they were removed by the action of electrical sparks.

6th. *Epilepsy.* In several instances of persevering application, not one successful case occurred.

7th. *Deafness.* Sparks thrown on the *Mastoid process*, and round the *Meatus auditorius externus*, and drawn from the same parts on the opposite side, usually afford relief; and about one in five are permanently cured.

8th. *Opacity of the Cornea.* This is sometimes cured by the long continued action of electricity thrown for ten minutes a day on the eye by a wooden point. When caused by the small-pox, it is said to yield most readily. I have known an instance in which considerable benefit was received from the application of electricity; but its use could not afterwards be discontinued for more than a week at a time, without a return of the disorder.

9th. *Gutta Serena.* The method of electrifying for opacity of the cornea has been successful in some instances of gutta serena; but there are very many unsuccessful cases.

10th. *Amenorrhœa.* Cases of suppressed menstruation are generally relieved by sparks and slight shocks; but in retention of the menses electricity has been tried without success.

11th. *Knee Cases.* In instances of pain and swelling of the knee the application of sparks has been effectual in about one case in ten.

12th. *Chronic Rheumatism.* Very numerous are the instances of success; the usual application is by sparks, for 10 or 15 minutes every day. In recent cases, a few days is sometimes sufficient; but in those of long standing, very considerable perseverance is often required.

13th. *Acute Rheumatism.* In one case out of six a cure was effected in about a month by the application of the electrified current of air from a point.

14th. *Palsy.* Moderate shocks, with sparks occasionally, have been successful in about one case of paralysis in every fourteen that have been tried.

St. Vitus's Dance has also been frequently relieved by electricity. There is indeed scarcely any disease in which some successful instances of its application are not recorded; but we are still in want of a scientific examination of the statements that have been made on this subject.

The nerves appear to be most powerfully affected by electricity; and the consequences of an electric charge sent through any part of the body are generally most conspicuous in their track. When the charge of a battery is sent through the head of a bird, its optic nerve is always injured or destroyed; and a similar shock, given to a larger animal, is said to produce a general prostration of strength, with trembling and depression. I once accidentally received a considerable charge from a battery through the head; the sensation was that of a violent but universal blow, followed by a transient loss of memory and indistinctness of vision, but no permanent injury ensued.

Mr. Morgan has stated, that if the diaphragm be brought into the circuit of a coated surface equal to two feet, fully charged, the lungs make a sudden effort, which is followed by a loud shout; but that if the charge be small, it never fails to produce a violent fit of laughter; and that even those whose calmness and solemnity are never disturbed by ludicrous occurrences, are rarely able to withstand the comic powers of electricity. The first effect of a strong charge on the diaphragm is frequently followed by involuntary sighs and tears, and sometimes by a fainting fit.

If the charge is passed through the spine, it produces a degree of incapacity in the lower extremities; so that if a person be standing at the time, he sometimes drops on his knees, or falls prostrate on the floor.

Great caution is required against the indiscriminate application of the shock, which has frequently produced unpleasant consequences when injudiciously applied. Its employment as a source of amusement in large companies should therefore be conducted with care; no evil effects are likely to occur by passing the charge through the arms.

Some experiments have been made to ascertain the supposed influence of electricity in the promotion of vegetation and animal life; but the results obtained by different inquirers are very contradictory, and it does not appear that any real progress has been made in this investigation, which certainly offers a most interesting object for the researches of the electrical philosopher.

One of the most interesting discoveries in Natural History was that of the electrical power of certain fishes which had been long known to possess the faculty of communicating at pleasure a tremor or benumbing sensation. Two of these animals, the Torpedo and the Gymnotus, are sufficiently well known; and a third the Silurus Electricus, has been described by Broussonet, under the name of Trembleur, in the Hist. de l'Academie Royale des Sciences, for 1782. It is but very imperfectly known.

The Torpedo was noticed by the ancient writers on natural history, and the analogous properties of the Gymnotus were observed towards the close of the 17th century; but the first demonstration of the identity of their powers with electricity, was effected by Mr. Walsh in 1772.* The shocks of these animals are com-

* See the 63d vol. of the Phil. Trans. p. 461, and following.

municated through all the substances that are conductors of electricity, and they are not transmitted through non-conductors; and those of the larger Gymnoti, when passed through a minute interruption in a metallic circuit, even produce a spark.

The dissection of these animals displays a peculiar organ, consisting of an extensive series of irregular columns divided by horizontal partitions, and exposing very considerable surface: the interstices appear to contain a fluid, and it is highly probable that the occasional propulsion of this fluid into the interstices, by which an extensive contact of two dissimilar bodies is suddenly effected, may be the source of their electric power.

The Torpedo is of the order of rays, it inhabits the Mediterranean, and the North seas; they rarely exceed eighteen or twenty pounds weight, when fully grown; the rapidity with which their shocks are communicated is considerable, fifty having been received in a minute and a half. The shocks appear to depend on the will of the animal, and their communication is constantly attended by a depression of its eyes; they are said to be four times as strong when the fish is insulated and surrounded by air.

Spallanzani has stated, that when dying, the Torpedo communicates its shocks more frequently than usual, but that they are then considerably weaker. He also asserts, that the young Torpedo can exercise this power the moment after its birth.

The Gymnotus Electricus, or Surinam Eel, abounds in the rivers of Surinam and Senegal; it resembles a large eel, but is thicker in proportion to its length, and rather of an unpleasant appearance. Its usual length is about three feet; but it is said they occasionally occur of ten or twenty feet long, and of sufficient electric power to occasion the death of a human being. The electric organ is somewhat more simple than that of the Torpedo, but of very considerable extent.*

The nerves connected with the electric organs of both these animals, are much larger than those appropriated to any other part of the body.

The demonstration of the electrical origin

* See Mr. Hunter's account of the Gymnotus. Phil. Trans. vol. lxv. And of the Torpedo, vol. lxiii.

of the power of these fishes excited considerable attention; their relation to common electrical phenomena was shewn by Mr. Cavendish, in an admirable paper published in the 64th volume of the Philosophical Transactions. We have now abundance of facts that are strikingly analogous. The appearance of any definite quantity of electricity, and its tendency to escape, may be almost infinitely modified by disposing it on different conductors. A coated plate of Muscovy talc will appear scarcely electrified when it has received a considerable charge; and the jars in a battery may be multiplied to such extent, that a very trivial spark from them shall melt a considerable length of wire. The *intensity* of the electric power of the torpedo, and the Gymnotus, is so inconsiderable, that it will not penetrate any evident interval of air; but the multiplied surface of their electrical organs renders the *quantity* accumulated subservient to the destruction of their prey.

The actual proof of the active exertion of electric powers in the animal system afforded by this discovery, increased the speculations on the probability of its universal agency. Numerous were the hypotheses formed, and the conjectures advanced, but for a considerable period they excited no particular attention. In the year 1790, L. Galvani, professor of anatomy at Bologna, accidentally discovered that the passage of a small quantity of electricity through the nerve of a frog that had been recently killed, had the property of exciting distinct muscular contractions. He produced the same effect with atmospherical electricity; and afterwards by the mere contact of two different metals. His discoveries were published in 1791: he proved the phenomena to be electrical, and says, "if you lay bare the sciatic nerve of a frog, and remove the integuments, then place the nerve on a piece of zinc, and a muscle on a plate of gold, and connect these metals by any conducting substance, contractions are produced; but if non-conductors are used to connect the metals, contractions are not excited." The experiments of Galvani received considerable attention; they were varied and extended with the greatest perseverance and address by professor Volta, Dr. Valli, Humbolt, Fowler, Monro, Robison, and many others. Many curious facts resulted from these researches, but they are much too extensive to detail in this place; an account of them will be found in the supplement to the Encyclopedia Britannica, article Galvanism.*

The effects obtained in the experiments of these naturalists may be illustrated by very simple experiments. The most important facts they establish are, first, that the passage of a small quantity of electricity through the nerve or nerves of any animal, occasions a tremulous motion or contraction of the contiguous muscles, and sometimes an extension of the limbs. This effect takes place both in living animals and such as have been recently killed, and even in the detached limbs of these last. It is produced when the transmitted electricity is considerably too weak to affect the most delicate electrometer, and obtains in all animals for some time after death; their susceptibility being greatest at first, and gradually diminishing as the limbs stiffen. Animals with cold blood, as frogs and fishes, retain the power of action after death longer than others, sometimes for many hours or even days.

* See also Fowler's Essay on Animal Electricity; Valli's Experiments on Animal Electricity; and Cavallo's complete Treatise on Electricity, vol. iii. fourth edition.

Secondly. The same effects that are produced by the passage of electricity, also result from the contact of different metals with the nerves and muscles. If a communication be formed between any nerve and muscle by a single metal, contractions are but rarely produced, and when they appear are very feeble; but if two metals are employed in contact with each other, motion is always obtained, and the effects are most considerable when the metals are most essentially different; thus zinc and gold, or zinc and silver, form a very active combination.

Thirdly. By the same means that muscular motion is excited in these trials, some of the senses are remarkably affected, as will be evident when the experiment is made on living animals.

The demonstration of these facts is easily effected. For the excitation of muscular motion any small animals may be employed; the most convenient are frogs and fishes. Frogs are peculiarly susceptible. If one of these animals be employed alive, a piece of tinfoil may be pasted on its back, and the frog being then placed on a plate of zinc, spasmodic convulsions will be produced whenever a communication is made by a wire between the zinc and the tinfoil. This experiment will succeed either in the open air or under water.

Small flounders, which may be usually obtained alive at the fishmongers, are also convenient for these experiments. The flounder is to be placed in a dish upon a slip of zinc, a shilling is to be placed upon its back, and whenever the zinc and the shilling are brought into metallic communication by means of a wire, strong muscular contractions are produced.

The smallest charge from a Leyden phial, (even such as will not produce a spark), when passed through a frog that has been recently killed, will produce muscular motion in it.

The most convenient preparation of the limbs of a frog is obtained by separating the head and upper extremities from the rest of the body, and removing the skin and the contents of the abdomen from the lower extremity; the crural nerves may then be distinctly seen, and the spine may be separated below their insertion into it, and will then remain attached to the legs by the nerves only. All the superfluous part of the spine is to be cut off, and the small piece that remains attached to the nerve is to be wrapped round with tinfoil. With this preparation various experiments may be made; the following are as illustrative as any.

Place two small glasses full of water near each other with the legs of the frog in one and the spine coated with tinfoil in the other; connect the two glasses by a silver wire, the legs will move, and sometimes so powerfully as to quit the glass.

Hold the prepared frog by one leg, the other hanging down with the coated piece of spine in contact with it. Interpose a piece of silver (a dollar or other coin,) between the lower thigh and the nerves, so that it may form a communication on the one side with the thigh, and on the other with the coated nerves; the hanging leg will immediately vibrate very powerfully.

Experiments of this kind have been made on almost every description of animals, from the grashopper to the ox, and the effects of contraction or motion, occur in all even in those that have been considered destitute of nerves Place a dollar upon a large plate of zinc, and put a leech upon the dollar; so long as it is in contact with the silver only, it will evince no uneasiness; but if in moving about it should stretch beyond the silver, and come in contact with the zinc, it will immediately recoil as if from a sudden pang.

Let any one place a piece of silver *upon* the tongue, and a piece of zinc *under* the tongue, or vice versa. Whilst the metals remain separate no effect is perceived, but if their edges are brought in contact, a slight shock will be felt, and a peculiar taste be experienced; occasionally, also, when the surfaces of the metals are extensive, a bright flash of light appears to pass before the eyes. This latter effect may be also produced by placing one metal between the upper lip and the gums, and the other upon the tongue, and then effecting the contact between them, or by covering the bulb of the eye with tinfoil, placing a silver spoon in the mouth, and then completing a metallic communication between the tinfoil and the spoon.

These effects, produced by the contact of dissimilar metals evince the excitation of some power by that means, and destroy the hypothesis proposed by Galvani to account for the muscular contractions he had produced.

Galvani supposed, that different parts of an

animal are naturally or by some process of nature in opposite states of electricity, and that the contractions are produced by effecting a metallic communication between them. Professor Volta opposed this statement, and shewed that the effect depended on the contact of dissimilar bodies, and not upon a communication between different parts of the animal. He demonstrated, that contractions might be excited in either part singly, by the application of two different metals or other very dissimilar substances, and accounted for the phenomena, by assigning a principle of electro-motion, (or power of producing a circulation of electricity,) to any circle formed by three bodies of different conducting powers. Thus zinc, silver, and the moisture of the animal are three bodies of different conducting powers, and they produce in these experiments the same effect as artificial electricity. He found also, that the effects were produced by a single metal and two different fluids, as well as by a single fluid and two different metals, and thus accounted for the effect obtained by a single metal when connected with opposite parts of an animal.

Alkaline sulphurets, (liver of sulphur,) and

silver, he found as effectual as zinc and silver, and he demonstrated, that either of these combinations must form a circle with the animal before they can act upon it. A demonstration of this fact is easily obtained. Place a cup of silver filled with water on a plate of zinc standing upon a table, and touch the water with the tip of the tongue; no particular sensation will be felt, for the body does not then form a circle with the metals. Moisten the hands, and grasp the plate of zinc with them, whilst the tongue is brought to touch the water; a peculiar sensation, and a saline taste will be immediately experienced; for the body then forms a communication between the opposite surfaces of the associated metals; which is the condition established by the experiments of Volta.

The peculiar action exerted by combinations of this kind appears to be instrumental in the production of some effects which have been often observed, but were inexplicable before the discovery of these facts.

It has been often noticed, by those who drink porter, that that fluid has a materially different flavour when drank out of a pewter pot, than when drank from a vessel of glass or earthen ware. In this case the moisture of the lips and the porter, are two different fluids between which a metallic connexion is formed by the pewter of the pot, and this it has been seen is a proper voltaic circle. Professor Robison has asserted, that even the flavour of a pinch of snuff is affected by keeping it in a box of tinned iron, from the surface of which a part of the tin has been worn by long use.

The sheathing of ships cannot be securely fastened by iron bolts; for the copper sheathing, the iron bolts, and the sea-water, form a Voltaic combination; and the metals are soon corroded at their juncture: copper bolts are now therefore generally employed.

Vessels that are soldered, tarnish soonest at the seams, for there two metals form a voltaic circuit with the water of the atmosphere.

On the same principle, the Etruscan inscriptions engraved upon pure lead are preserved to the present time, whilst medals of mixed metals of a much more recent date are much corroded. There are many other circumstances of a similar nature, which may be easily comprehended when attentively considered and compared with these.

Volta supposed that the faculty of forming these combinations depended on the different conducting powers of the associated bodies, and he divided them into classes, according to their fitness for the production of these effects.—There are two principal classes; 1st. Dry and perfect conductors, as metals and charcoal. 2d. Imperfect conductors, as fluids and fibrous solids, which derive their conducting power from the fluids they contain.

A proper voltaic combination consists of three bodies taken from these two classes, and their energies are greater in proportion as they differ from each other more considerably.

When two perfect conductors are combined with one imperfect conductor, (as silver and zinc with water,) the combination is said to be of the first order. When two imperfect conductors are combined with one perfect conductor, (as silver with alkaline sulphuret, and water or acids,) the combination is said to be of the second order.

Sir Humphrey Davy has constructed the following tables, expressive of some simple combinations of each kind.

FIRST ORDER.

Most oxidable Substances.	Less oxidabl Substances.	Oxydating Fluids.
Zinc.......	With gold, charcoal, silver, copper, tin, iron, mercury.	Solutions of nitric acid in water, of muriatic acid, of sulphuric acid, &c. Water holding in solution oxygen, atmospheric air, &c.
Iron	.... gold, charcoal, silver, copper, tin.	
Tin........	.... gold, silver, charcoal.	
Lead.......	.... gold, silver.	
Copper.....	.... gold, silver.	Solution of nitrate of silver and mercury. Nitric acid, acetous acid.
Silver......	.... gold.	..Nitric acid.

SECOND ORDER.

Perfect Conductors.	Imperfect Conductors.	Imperfect Conductors.
Charcoal ... Copper .... Silver ...... Lead ...... Tin........ Iron .... Zinc.......	Solutions of alkaline hydro-sulphurets, capable of acting on the first three metals, but not on the last.	Solutions of nitrous acid, chlorine, muriatic acid, &c. capable of acting on all the metals.

These combinations are more or less powerful, nearly in the order of their arrangement, the most active occupying the top of the columns.

Some difficulty attends the demonstration of the electrometrical effects which, agreeable to the supposition of Volta, should take place with these combinations. That excellent electrician succeeded in producing them by the aid of the condenser, which has been already described as his invention. He proved, that when two metals are employed, the humid or imperfect conductor combined with them has but a very trivial share in the production of Galvanic effects, the metals themselves being the primary source of electrical motion. When two fluids and a metal are employed, one of the fluids only acts with the metal as a motor of electricity, the other serving merely to facilitate the effect by its conducting power, or to convey a current of electricity from one of the motors to the other.

The endeavours of this celebrated philosopher, to establish these principles, led him to attempt the arrangement of more powerful combinations, and the concentration of their effects; and in the year 1800, he communicated in a letter to Sir Joseph Banks, a description of some arrangements by which these purposes were attained, and very remarkable electrical powers manifested.

This communication must be regarded as the first dawn of a splendid era in electrical philosophy. which has been advanced by it from the glimmer of twilight to the unclouded brilliance of open day. It was hailed by philosophers with an enthusiasm commensurate to its importance, and employed with a degree of skill, attention, and assiduity, as unprecedented as the success by which it has been attended.

Amongst the most active inquirers may be enumerated Messrs. Nicholson and Carlisle, Mr. Cruickshanks, Dr. Henry, Sir H. Davy, Dr. Wollaston, Messrs. Pepys, Sylvester, Children, Ritter, De Luc, Pfaff, Thenard, Van Marum, Biot, Desormes, Priestley, Bostock, Simon, Wilkinson, Hisinger, Cuthbertson, and Berzelius. The detail of their labours would occupy volumes; the apportionment of their praise will be the duty and the pride of future ages.

The arrangements proposed by Volta are

named, in just commemoration of their inventor, the Voltaic apparatus; and the electrical effects they produce are considered by the appellation of Voltaic electricity. This subject will be detailed systematically in the following section.

PART IV.

VOLTAIC ELECTRICITY.

CHAP. I.

Structure of the Voltaic Apparatus, and Nature of its Electrical Phenomena.

In the enumeration of simple Voltaic combinations, it has been stated that those of the first order consist of two metals and a fluid, and that Signior Volta supposes the association of the two metals to be the primary cause of the phenomena they produce.

There are two methods by which the production of the opposite states of electricity, by the contact of dissimilar metals, may be exhibited. The first requires for its action on the electrometer, the aid of a condenser; the second is more simple, and produces its action immediately on the electrometer.

The most convenient condenser has been

already described;* when employed in these experiments its insulated plate is to be connected with a very delicate gold leaf electrometer.

Experiment 1. Procure two circular plates, about four inches diameter, the one of copper, and the other of zinc, perfectly clean and bright; let an insulating handle be screwed into the centre of each plate. Hold the plates by their insulating handles and apply their flat surfaces together, suffering them to remain in contact about a second, then separate them and touch the insulated plate of the condenser with the copper; bring the zinc and copper in contact with each other again, then separate them and touch the condenser with the copper; repeat this operation ten or twelve times, then remove the uninsulated plate of the condenser, and the electrometer will diverge with *negative* electricity.

Experiment 2. Take off the electricity of the electrometer, and prepare the condenser as before. Repeat the contacts of the insulated zinc and copper plates, and every time they are separated touch the condenser with the zinc plate.

* See page 128; the instrument is represented by Fig 19.

Remove the insulated plate of the condenser, and the electrometer will diverge with *positive* electricity.

Experiment 3. Place a copper plate upon a table, and a zinc plate upon the copper. Lay a disk of moistened leather, pasteboard, or cloth upon the zinc, and connect this moist conductor by means of a wire with the insulated plate of the condenser. After about half a minute's contact the wire may be removed, and the insulated disk of the condenser being connected with the cap of a condensing electrometer, (Fig. 20.) its uninsulated disk is to be turned back: this process transfers the charge of the large condenser to the cap of the electrometer. The condenser is now to be removed, and the small uninsulated plate of the electrometer being turned back, its gold leaves will diverge very slightly with *positive* electricity. If the group of copper, zinc, and wetted cloth, be then reversed, and the contact of the condenser be established with the copper, on transferring the electricity of the condenser to the electrometer its leaves will diverge *negatively*.*

* These experiments require great care. The electrometer and the insulated disk of the condenser should be very perfectly

Experiment 4. Procure a small sieve of copper, made by drilling a number of small holes in a concave piece of that metal. Fit an insulating handle to the sieve, and fill it with zinc filings. Place a broad plate of tin or brass on the cap of a gold leaf electrometer. Hold the sieve by its insulating handle, and sift the zinc filings through it upon the electrometer: the leaves will diverge with *positive* electricity; and if the copper sieve be examined it will be found *negatively* electrified.

Experiment 5. Repeat the preceding experiment, substituting a zinc sieve and copper filings, for the copper sieve and zinc filings. The sieve will be electrified *positively*, and the filings *negatively*.

Similar experiments may be made with other metals; almost any two that are dissimilar become oppositely electrified when they are brought in contact with each other, and afterwards separated. If we enumerate them in the following order, namely, zinc, iron, tin, lead, copper, silver, gold, platina, it will be found, on trial, that

insulated, and the manipulations be conducted with great care and attention. Even when all these circumstances are observed, the divergence of the electrometer is very slight.

any one of these becomes *positive* by contact with any that *follow* it, and *negative* by contact with any that *precede* it.

From these facts it appears probable, that when two different metals are associated together, their natural attraction for the electric fluid is altered, and a portion consequently flows from one to the other. If this be admitted, the manifestation of the opposite states of electricity, when the metals are separated, is analogous to the usual effects of electrical excitation.

Copper and zinc are the metals most usually employed in the construction of Voltaic apparatus, for their effects are greater, in proportion to the value of the metals, than those of any other combination. Silver and zinc, or gold and zinc, would be more powerful, but not so much so as to compensate for the increased expense.

As the effects produced by a single pair of metals, of any size, are still exceedingly feeble, attempts were made to combine the action of several pairs. Professor Robison arranged a series of zinc and silver plates, about the size of a shilling, so as to form a rouleau; and on applying his tongue to the edge of this, the sensation experienced was more manifest than by a simple pair of metals; but its power in other respects did not appear more considerable. In this arrangement every zinc plate was necessarily between two silver plates, and every silver plate between two of zinc, with the exception of the first and last. Now it has been stated, that the contact of zinc with silver, or copper, occasions some electric fluid to flow from either of those metals to it; and, consequently, when a single pair of metals are associated, the outer surface of the zinc appears positive, and that of the silver or copper negative. But if both surfaces of the zinc are in contact with copper or silver, electricity will flow into it in contrary directions, so that neither surface can exhibit the effect; and the same circumstance occurs, in a contrary order, when both surfaces of a silver or copper plate are in contact with zinc. Hence every arrangement of this kind, however numerous the pairs of metal, will exhibit at its opposite extremities the powers of a single pair of metals only.

Volta had the penetration to ascertain the cause of this defect in the rouleau of Professor

Robison; and his ingenuity supplied a means of obviating it. His experiments on the combination of two metals with an imperfect conductor, (as water or saline fluids,) had taught him that the electro-motive power of these fluids interfered but little with the more powerful energy of the combined metals; and, that in fact they acted principally as conductors to that energy. He therefore interposed imperfect conductors of this kind between a series of pairs of metal, and thus combined their power without producing a counteracting current; for the zinc and silver, or zinc and copper, were then in contact with each other at one surface only, but the conducting communication existed throughout.

To construct an apparatus of this kind, procure a number of plates of zinc and copper, or zinc and silver, either round or square, of any size; and an equal number of pieces of cloth, leather, or pasteboard, of the same form, but rather smaller. Soak these last in salt water, until they are thoroughly moistened; place a plate of silver, (or copper,) upon the table, then upon that place a plate of zinc, and on the zinc one of the moistened disks; upon this a second series of silver, zinc, and moistened cloth, (or pasteboard,) in the same order; and thus consecutively until a series of fifty or sixty repetitions have been placed one upon the other.* Particular care must be taken to place the plates in regular order; if in the first group silver is placed lowest, zinc next, and then the moistened disk, the same disposition must be observed throughout.

Experiment 6. The Voltaic pile being thus formed, let the operator moisten both his hands with brine, and grasp a silver spoon in each. If the top of the pile be then touched with one spoon, and the bottom with the other, a distinct but slight shock will be felt at every repetition of the contacts. This shock resembles very nearly the sensation produced by a very large electrical battery weakly charged; it is greater in proportion to the number of groups of which the pile is composed. If the communication is made with any part of the face near

* To prevent the pieces from falling down, when their number is considerable, it is usual to build them up between three pillars of varnished glass, placed at equal distances from each other in a triangle, and cemented into a thick piece of wood, which serves as a base for the pile. See Fig. 36.

the eyes, or with a silver spoon held in the mouth, a vivid flash of light is perceived at the moment of contact, and that whether the eyes be open or shut.

The power of an apparatus of this kind continues for some time, but gradually diminishes, the zinc surfaces becoming oxidated by the action of the moisture; it therefore requires to be taken to pieces and cleaned, an operation that is very troublesome when the number of plates is considerable. This inconvenience was diminished by soldering each pair of zinc and copper plates together, instead of simply laying them on each other; and a further improvement was devised by Mr. Cruickshanks, which consisted in cementing the pairs of plates in regular order, in grooves made in the sides of a mahogany trough, so as to form water-tight cells between each pair. These cells being filled with water, or any conducting fluid, served as a substitute for the moistened disks used in the pile; and as the fluid could be easily poured out, and replaced, it required considerably less time to keep it in proper order. This form of the apparatus, which is called the Voltaic trough, or battery, has been much used in this country;

it is perhaps, on the whole, the best arrangement hitherto devised, and its construction is sufficiently simple.

The zinc plates are made by casting that metal in an iron or brass mould; they may be about an eighth of an inch thick. The copper need not exceed twelve or fourteen ounces to the square foot, and may be soldered to the zinc at one edge only, the other three being secured by cement in the trough.

The trough must have as many grooves in its sides as the number of plates it is intended to contain, which should be fewer in proportion to their size, otherwise the apparatus will be inconvenient from its weight. When the plates are not more than three inches square, their number in one trough may be fifty, and the distance of the grooves from three-eighths to half an inch. The trough must be made of very dry wood, and put together with white-lead or cement. The plates being placed to the fire, the trough is to be well warmed, and placed horizontally on a level table, with its bottom downwards, very hot cement is then to be poured into it, until the bottom is covered to the depth of a quarter of an inch. During this process the plates will have become warm, and they are then to be quickly slided into the grooves and pushed firmly to the bottom, so as to bed themselves securely in the cement. In this way the plates are very perfectly cemented at the bottom, and when this cement is sufficiently cool, a slip of thin deal is to be slightly nailed on the top edge of one of the sides of the trough, so as to overhang the inner surface about a quarter of an inch. The trough being about three-quarters of an inch deeper than the diameter of the plates, there will be an interval between their top edges and the deal slip; and when the side of the trough to which the slip is attached is laid flat upon the table, this interval forms a channel into which very hot cement is to be poured, and it will flow between each pair of plates, so as to cement one side of all the cells perfectly. As soon as the channel is quite full of fluid cement, the strip of deal is to be torn off, and the trough inclined so as to admit the superfluous cement to run out. When this is effected and the cement cool, a slip of deal is to be nailed on the opposite side, and the same process pursued with that. The instrument will then be cemented in the most perfect manner and it may be cleaned off and varnished.

I have been rather particular in this description, because I have not yet noticed an account of this instrument that would enable any one successfully to attempt its construction. It is represented by fig. 37.

Volta proposed another form of the apparatus, to which he gave the name of "couronne des tasses." It consists of a row of wine glasses, or cups, containing salt and water, or any saline fluid. Into each of these one extremity of a metallic arc, consisting of a plate of zinc connected by a wire with a plate of copper, is plunged. These arcs are so arranged that the copper extremity of the first is in the same glass with the zinc extremity of the second, the copper of the second with the zinc of the third, and so on in regular order through the whole series, the extreme glasses forming the opposite extremities. An apparatus of this kind occupies considerable space; it has been modified of late by employing troughs of mahogany, divided into cells by glass partitions; or troughs of Wedgewood ware, with the partitions formed of the same material: the plates of copper and zinc are soldered together in one point only, and each pair of plates is arranged so as to enclose a partition between them; there is consequently in each cell a copper plate connected with the zinc of an adjoining cell, and so on in regular succession. The troughs usually contain ten or twelve pairs of plates, and these are connected together by a slip of baked wood, so that they may be lifted into or out of the cells together. See Fig. 38.

This construction has the convenience of admitting the fluid to remain in the trough whilst the action is suspended by lifting the plates from the cells; the plates are also easily replaced when injured or worn; but both surfaces being exposed they wear much faster, and it does not appear that any effect commensurate to the increased surface is ever obtained by them.

The size of the plates in the Voltaic apparatus has been greatly varied; they have been constructed as small as half an inch diameter, and as large as 2 feet 8 inches by 6 feet. The largest plates are useful only for some particular purposes, to be hereafter described. The most useful sizes are from 2 inches to 6 inches square.

It is obvious that, in all the arrangements described, the order of the plates is similar, the copper and zinc alternating regularly with each other; hence, if the first plate in any battery be zinc, the last will be copper; and in all cases the uncombined surfaces of the copper and the zinc plates are opposite to each other. That end of the battery to which all the copper surfaces are turned is called the copper extremity; that to which the zinc surfaces incline is called the zinc extremity.

When many plates are required, the power of several batteries may be combined together by connecting them in proper order, endwise, with slips of copper: the usual rule is to connect the zinc end of one battery with the copper end of another, and so on; for in this way their plates tend all in one direction: if one battery, or even a few plates in an extensive series, should be reversed, a very considerable diminution of power will be sustained.

The fluid interposed between each pair of plates is essential to the combination of their power; but independent of this, it appears to have a manifest influence on the effects produced. A single pair of zinc and copper plates will not readily affect the common condensing electrometer, if they are merely placed in contact with each other, whilst one of them is connected with it; but if a piece of wet cloth is placed on the zinc, the electrometer is usually affected. Again, if a pile or battery of 50 pairs of plates be put together with water interposed, the shock will be exceedingly faint; but if the water have a considerable portion of salt dissolved in it, the shock will be much stronger. Hence the introduction of the interposed fluid is, with some propriety, called exciting the Voltaic battery.

If the opposite extremities of any excited Voltaic battery be accurately examined, they will be found in different states of electricity; a few plates will manifest this by the aid of the condenser; but with a series of fifty groups, a delicate gold leaf electrometer will be affected without the aid of the condenser. With one hundred pairs the divergence of the gold leaves is sufficiently distinct; and with a series of one thousand groups, even pith balls are made to diverge. In these experiments, a wire proceeding from one extremity of the battery is to be connected with the foot of the electrometer, whilst a wire proceeding from the opposite extremity is brought to touch its cap. The electricity of the zinc side is always positive; that of the copper side always negative.

It is a singular fact, that though the shock from the Voltaic battery is increased by brine, or other fluids that have some chemical action on the zinc, this is by no means the case with its electrical indications. Volta first proved this by noting the divergence of an electrometer produced by his "couronne des tasses," when charged with water, and then putting a pinch of salt into each cup, which increased the shock, but not the effect on the electrometer. I have made many experiments of this kind on an extensive scale, employing from one hundred to one thousand groups, and have found the electrical effects greatest when the chemical action has been least. One hundred plates were charged with water, and their effect on the electrometer noted; the shock was scarcely appreciable. The water was poured out, and a weak solution of muriatic acid introduced; the shock was greatly increased, but the effect on the electrometer was evidently diminished. I have varied and extended these experiments with some care, and have constantly found that the most considerable electrical effects are produced when common river water is employed; but the most remarkable circumstance is, that in this way of employing a battery, the most distinct evidence may be obtained of a real increase of the electrical effects of the Voltaic apparatus, by combining with it an electrical battery. If a wire connected with one extremity of a Voltaic battery charged with water, be brought to the other extremity, a very faint spark only will be perceived, even if the series be 800 or 1000; and if it extend to but one hundred, there is rarely any perceptible effect. But if a wire proceeding from each extremity, be respectively connected with the inner and outer surface of an electrical battery, of not less than twelve square feet of coated surface, this will be charged so rapidly, that sparks may be obtained from it in rapid succession, by connecting a fine iron wire with its outer surface, and successively striking the knob with its other extremity: these sparks are so strong when the charge is communicated by a series of three or four hundred, that the end of the iron wire is made to scintillate or throw off sparks; and with a series of one thousand, the sparks are attended by a distinct crackling noise, and have sufficient power to burn thin metallic leaves, though no such power is possessed by the Voltaic battery itself, when employed with river water.

On one occasion I employed in this way four hundred pairs of four inch plates; they affected an electrometer distinctly, but produced no action on combustible bodies. I then proceeded to interpose an electrical battery between the opposite extremities of the Voltaic apparatus; it was charged instantly, so as to produce scintillating sparks in quick succession, and the charge was so incessantly kept up, that almost continuous discharges were procured; and with these phosphorus was inflamed, and fulminating mercury exploded.

The most powerful electrical machine would not produce an effect of this kind, though by its long continued action the battery might be charged to a much higher degree. The quantity of electricity put in motion by the Voltaic apparatus must therefore be very considerable;

but from the circumstances of its production, a very limited intensity is all that can be obtained.

There is reason to believe that a considerable portion of the effect produced by a Voltaic combination, is lost by the conducting power of the bodies with which it is connected. When an electrical battery is interposed between its poles, there is a kind of reservoir for the reception of the power excited; this will therefore accumulate in proportion to the extent of that reservoir; and it is probably for this reason, that more considerable effects are manifested by a large battery than by a single jar, more from a large jar than from a small one, and more from this last, than from the apparatus itself.

A single jar is always charged by the shortest possible contact with a Voltaic apparatus excited with water, to rather more than the intensity of the apparatus itself, and will consequently affect an electrometer somewhat more distinctly.

The action on the electrometer increases in every Voltaic apparatus with the number of its pairs of plates, and Volta has stated that the increase is exactly proportioned to the number. According to Ritter, the effect of the shock, when water is used, increases with the number of plates to the extent of five or six hundred; it then decreases with higher numbers. When the exciting fluid is brine, the effect increases with the number, and continues a proportionate increase with the most extensive series that have been hitherto tried. I have observed the same effect with weak acid solutions; the shock from one thousand plates is very sharp and painful.

The power of a very limited Voltaic battery to produce muscular motion is considerable; if two wires be placed in the ears of an ox,* a sheep, or other animal, soon after its death, and one of the wires be connected with one extremity of a battery of one hundred plates, whilst the other wire is brought occasionally in contact with the opposite extremity, at every com-

* The parts with which the wires are brought in contact must be previously wetted with salt water, the intensity of Voltaic electricity being insufficient to overcome the resistance of the dry cuticle. This circumstance is also necessary to be attended to when Voltaic electricity is applied to the cure of disease.

pletion of the circuit very strong muscular action will be excited, the eyes may be made to move, and the apparent effects of smelling, chewing, &c. may be produced. With a much less power the legs of a frog may be made to move very powerfully, and even to leap to a considerable distance, and that too an hour after its death.*

If a battery has been out of use for some time, it will, when charged with water, evince but a slight effect in the production of the shock; and however extensive it may be, will scarcely evolve a perceptible spark, but its ac-

* The tongue of an ox was secured to a table by an iron skewer; when the power of a Voltaic battery was applied, the tongue was drawn in with such force as to detach the skewer from the table. An entire sheep exhibited motions resembling the struggles of animals in an epileptic state, but more powerful than the natural actions. ("Wilkinson's Elements," vol. ii. p. 464.) It was supposed that the application of the Voltaic apparatus might be serviceable in cases of suspended animation; and trials have been actually made on some criminals a short time after their execution. Very violent muscular action was produced, but no evidence of returning life. (See Aldini on Galvanism, p 191.) This power has been also applied to the cure of disease, and apparently with some success; but the cases hitherto published are scarcely of sufficient importance to warrant any amplification on this subject.

tion on the electrometer, and its power of charging an electrical battery, will be found considerable. If a weak acid solution be now substituted for the water, the effect of the shock will be remarkably increased, and a vivid spark may be obtained; but if the electrometer be applied, its divergence will be less than in the preceding instance, and the same may be said of the power to charge a battery. If the acid be now poured out, and the troughs washed, and refilled with water, the force of the shock will remain, or experience a slight increase; the spark will be weaker, and the electrical effects a little stronger, than with the acid charge: but the usual effect of the excitation by water will not be correctly obtained until the apparatus has been frequently rinsed out and refilled.

From the preceding facts it is evident, that, for the illustration of the electrical effects of the Voltaic apparatus, a very extensive series of plates, excited by water only, will be the most efficient arrangement. Such an apparatus has the capital advantage of maintaining its power without any renewed attention for months, and probably for years. Its energies are increased when communicated to an electrical battery; and as the intensity of the charge it communicates increases with the number of plates through all the series hitherto tried, there is little doubt that if 50 or 100,000 plates were employed, a considerable charge might be constantly kept up in an electrical battery, and that at no expense but the first cost of the apparatus; which would be sufficient if formed of plates two inches square.

CHAP. II.

On the Chemical Effects of the Voltaic Apparatus.

THE chemical phenomena produced by Voltaic electricity are much more remarkable and extensive than those that result from the action of the ordinary electrical apparatus. Many of them are produced by the most simple combinations, and are conducted with a degree of tranquillity analogous to the spontaneous operations of nature.

Mechanical action is generally evident during the operations of common electricity, but such phenomena are rarely exhibited by the Voltaic apparatus. Few are the instances in which the action of the electrical machine is unattended by the appearance of light, yet it rarely exhibits any unequivocal effect of heat, but what may be considered as the consequence of its mechanical agency.

In the Voltaic apparatus, when no light is evolved, an elevation of temperature may be usually observed; and when, by its intense action, there is a copious evolution of light, heat is produced in a superior degree to that which results from any other process of art.

The chemical agency of the Voltaic apparatus was discovered by Messrs. Carlisle and Nicholson, during the first experiments made with it in this country;* and within a year after its introduction, the distinct peculiarities of its action, in this way, were partly developed by the activity of the British philosophers.

The decompositions produced by the Voltaic apparatus are effected with remarkable precision. The component parts of the bodies subjected to its action are separated at some distance from each other, and no observable change occurs in the intermediate space. If two wires of gold or platina, for instance, are respectively connected with the opposite extremities of a Voltaic battery, and are then plunged at some distance from each other into a vessel of water, bubbles of air will soon arise from each, but in the greatest quantity from that connected with the copper (or negative) side of the battery. If these gases are collected, by suffering them to

* See Nicholson's Journal, 4to. vol. iv. p. 179, &c.

rise into two small tubes filled with water, and placed over the respective wires, that produced in the greatest quantity will be found on examination to be hydrogen, and that produced in the smallest quantity oxygen; their relative quantities being by bulk nearly as one to two, which is the proportion in which they are found by experiment to combine, and form water.

If, instead of gold or platina wires, any metals more readily susceptible of oxydation be employed, the wire connected with the zinc (or positive) extremity of the battery will be oxydated, and evolve no gas; but that connected with the copper (or negative) side will continue to evolve hydrogen.

Hence it is an established fact, that whenever water is made the medium of communication between two wires proceeding from the opposite surfaces of a Voltaic battery, oxygen is separated by the positive wire, and hydrogen by the negative.

The demonstration of these facts may be effected by a very simple apparatus. To evince the distinct appearance of the gases, two platina wires may be inserted through corks in the opposite ends of a glass tube filled with water.

On connecting the wires with the opposite ends of a Voltaic battery of 50 or 100 two-inch plates, gas will be copiously evolved from each; and the process will take place, though with diminished energy, when the ends of the wires are distant from each other even three feet!

When both gases are collected in one tube, they may be reconverted into water by passing an electrical spark through them. A simple and elegant arrangement for this experiment is shewn by Fig 39. It consists of a brass cup, supported by a thick brass wire, which passes through its bottom and rises about an inch within it. This wire serves as a support for a small receiver, having a platina wire sealed in its top and projecting withinside near half an inch. There is a small hole in the top of the thick brass wire, which receives the end of a thin wire of platina, whose length extends to within a twentieth of an inch of the extremity of that in the top of the small receiver. The apparatus is filled with water, and a connexion being then established between one extremity of a Voltaic battery and the brass cup; the circuit is completed by a wire brought from the opposite extremity to the platina at the top of the receiver. Gas is then evolved from both wires, and rises to the top of the receiver, depressing the water until it sinks below the end of the upper wire; the process then stops, and a spark may be conveyed through the collected gas by passing the charge of a small Leyden jar from one wire to the other, whilst the receiver is held firmly in its situation. The gas inflames, and the water rises to the top of the receiver. The Leyden jar is unnecessary when the Voltaic battery has sufficient power to produce the spark.

The most simple arrangement to collect the gases separately, is represented by Figure 40. A and B are two separate receivers, formed of glass tubes three or four inches long, having each a platina wire sealed in its top, and extending within it through its whole length. These receivers are to be filled with water, and inverted in a small glass half filled with that fluid. The receivers are then to be connected respectively with the opposite surfaces of a Voltaic battery. That connected with the copper surface will be soon filled with hydrogen, and that attached to the zinc surface will in the same time be half filled with oxygen.

This effect will even take place when each receiver is placed in a separate glass, provided the water in the glasses be connected by any moist fibrous substance; as a moistened thread or a piece of moistened asbestos. This curious fact, which was first noticed by Sir H. Davy, and has been extended by him to a most interesting series of experiments, renders the explanation of these phenomena very difficult. The oxygen and hydrogen procured, are supposed to result from the decomposition of the same particle of water, yet they appear at very distant situations; at the opposite extremities of a long tube, or even in separate glasses of water connected by a moistened fibre. At whatever part of the circuit the decomposition is effected, it therefore appears certain that one of the gases, after its separation from the other, must traverse a considerable portion of the fluid in an invisible state, which is lost the moment it reaches its correspondent wire!

Mr. Cruickshank discovered, that the same agency which evolves hydrogen from water will revive metals from their solution in acids, and produce the separation of alkaline and acid matter from neutral fluids; these phenomena may be illustrated by very simple experiments.

Experiment 7. Adapt two corks to a glass tube of half an inch diameter, and four inches long, with two wires passing through the corks to within an inch of each other. Fill the tube with a dilute solution of acetate of lead, and place it in the circuit of a Voltaic battery. Metallic lamina, and fibres will almost immediately appear adhering to the negative wire, and will soon cover it with a beautiful vegetation of metallic lead. This experiment may be repeated with muriate of tin, or nitrate of silver with nearly the same result; with tin the appearance is very beautiful. Many other metals are revived, but none that I have tried appear with the same metallic brilliance.

Experiment 8. Bend a small glass tube in the shape of the letter V, so that it may form an inverted syphon, and introduce a platina wire into each of its legs. Fill this syphon with neutralised infusion of red cabbage:* con-

* To prepare this, when intended as a delicate test for acid or alkali, minced leaves of red cabbage are to be infused for a short time in a sufficient quantity of warm distilled water to cover them; the fluid being strained off, will be found to have

nect one of the wires with the negative, and the other with the positive side of a Voltaic battery; gas will be evolved, and in a short time the liquor in the positive leg of the syphon will become red, and that in the negative leg, green. Reverse the connexions of the wires, so that the one which was positive may be rendered negative, and that which was negative become positive. The red liquor will first resume its original blue colour, and then become green; and the green liquor, after returning to its original blue, will become red.

These changes may be repeated at pleasure for a considerable time by changing the connexions of the wires; they may be effected with very moderate power, even thirty pairs of two inch plates.

This determination of alkali to the negative wire, and acid to the positive wire, was mis-

acquired a fine blue colour, which becomes green by the contact of alkalies, and red by acids; it cannot be preserved for any length of time. A more useful infusion for experiments of transfer is made by adding to every pint of water, poured on the minced leaves, a few drops of sulphuric acid; this extracts a red infusion, which is more readily preserved, and a portion of it may be neutralized at any time, by cautiously adding a few drops of ammonia, until the blue colour appears.

taken at first for an evidence of the production of these bodies by the Voltaic apparatus. Sir H. Davy investigated the phenomena with the most indefatigable industry and consummate skill; and demonstrated by a series of admirable experiments, that they arose from the operation of some peculiar power in the Voltaic apparatus, by which hydrogen, inflammable bodies, alkalies and metals, are attracted to its negative (or copper) surface, and oxygen, and acids, to its positive (or zinc) surface. He shewed that this attraction is exerted with sufficient force to separate these substances from their most intimate combinations, and to manifest their presence when they exist even in the smallest quantities. Thus distilled water exposed to the action of the Voltaic apparatus, in separate vessels of glass connected by moistened fibres, had been observed to evince the presence of alkaline and of acid matter; the new experiments proved that the alkali arose from a partial decomposition of the glass, and the acid from the combination of the nascent oxygen of the water, with the nitrogen of the atmosphere.* Acid and alkali were indeed ap-

* Phil. Trans. for 1807, p. 1 to 56.

parent in a slight degree when vessels of pure gold were employed with common distilled water, but it was afterwards found that such water always contains a minute portion of saline matter; and it was shewn, that when water is slowly distilled in a silver still, and decomposed in gold vessels, out of the contact of the air, no trace of either alkali or acid appears.

These experiments displayed the importance of the Voltaic battery as an instrument of analysis; for the elements of almost all the bodies subjected to its action, were separated and collected at the wires connected with its opposite surfaces.

Very numerous experiments were made on this subject; the apparatus employed consisted usually of two cups, sometimes of glass, but more frequently when great accuracy was required, of agate, or gold; the cups were connected together by a few fibres of moistened asbestos,* and respectively connected with the opposite surfaces of a Voltaic battery. If a portion of any saline compound was placed in each cup, and the action of the battery con-

* For ordinary experiments moistened cotton may be substituted for asbestos.

tinued for a sufficient time, all the alkaline matter was collected in the negative cup, and the acid matter in the positive. Thus, when the common Glaubers salt, which consists of sulphuric acid and soda, is placed in solution in the apparatus, after a few hours, the positive cup will be found to contain a solution of sulphuric acid, and the negative cup a solution of soda. The acid and the alkali must consequently have been transmitted in opposite directions through the moistened fibre, or rather through the water it contains. Similar experiments may be made with any neutro-saline compound.

Any compound solution may be placed in one cup, and distilled water in the other. If the cup containing the solution be made positive the acid will remain in it, and the other element of the compound be transferred to the negative cup. If it be made negative, the acid will be transferred, and the other element will remain. In this way insoluble earths, or even metals may be transferred.

A very pleasing experiment of transfer may be made with three cups placed side by side in a line, and connected together by moistened cotton. Sulphate of potash may be placed in the middle cup, and blue infusion of cabbage in each of the others. When this apparatus is placed in the Voltaic circuit, the outer cups being respectively connected with the opposite sides of the battery, the sulphuric acid will collect in the positive cup and render its blue infusion red, and potash will deposit in the opposite cup and tinge its blue contents green.

The vessels themselves may be formed of compact saline bodies, as sulphate of lime, sulphate of barytes, &c and being filled with distilled water, and connected by moist fibres, their elements will be gradually separated, and collect at the opposite wires, but considerable time is required for this purpose.

So powerful are these means of decomposition and transfer, that the elements of compound bodies may be conveyed through chemical menstrua for which they have a strong attraction. Thus, when three vessels were employed, sulphate of potash being placed in that connected with the negative side of the battery, a solution of ammonia (which has a strong attraction for sulphuric acid) in the middle vessel; and water in that connected with the positive side. The sulphuric acid passed from the negative cup through the ammonia, and collected in the positive cup. By a variation of this experiment, an acid being substituted for the ammonia, and the sulphate of potash being rendered positive; the potash was transferred through the acid to the negative surface; and the same result was obtained with many other salts. It failed only when the intermediate fluid formed an insoluble compound with the transmitted substance; as in the attempt to transmit barytes through sulphuric acid, or sulphuric acid through a solution of barytes.

The want of chemical action between the interposed menstrua and the transmitted bodies, appears to arise from some peculiar annihilation of energy during the process, which is perhaps also the cause of the invisible transmission of gas. For acids may be transmitted through delicate vegetable colours without affecting them; and such is also the case with alkalies. To illustrate this, let three glass cups be arranged as before described, connected with each other by moistened cotton, and introduced into the Voltaic circuit; the centre cup being filled with blue infusion of cabbage, the positive cup with pure water tinged with the same infusion, and the negative cup with sulphate of soda; a redness will soon be produced in the water of the positive cup, and it will shortly become strongly acid. Now, the acid thus collected must have passed through the middle vessel, but the infusion it contains will experience no change of colour. By altering the connexions of the outer cups with the surfaces of the battery, the soda may be transferred in the same way; it will be collected in the tinged water of the negative cup, and render it green, but no effect will be apparent in the intermediate infusion through which it has passed.

The singular phenomena attendant on these experiments, and the constant uniformity of their results, evince decisively the existence of some property of Voltaic electricity analogous to the usual operations of chemical attraction. The opposite surfaces of the battery appear to have a natural attraction for different elementary bodies; inflammable substances, alkalies, earths, and oxides, being constantly determined to the negative surface; and oxygen, chlorine, and acids, to the positive surface: now, if it be conceived that these phenomena are occasioned by electrical attraction, they can only be explained by supposing that the attracted substances have naturally a contrary electricity to that of the surface to which they are determined; and such a supposition can scarcely be entertained without admitting, that chemical and electrical attraction are identical, "or produced by the same power acting in one case on masses, and in the other on particles." The illustrious philosopher, to whose skill and perseverance we are indebted for the ample developement of these facts, advanced a variety of phenomena in support of this opinion, and displayed the same ingenuity and talent in the structure of an hypothesis, as in the discovery of important truths.

If water be interposed between the wires from a Voltaic battery, oxygen separates at the positive wire, and hydrogen at the negative.—It is therefore supposed that oxygen is naturally negative, and hydrogen naturally positive; they consequently attract each other and form water, which is neutral, the electricities compensating each other. Now the union of the oxygen and hydrogen arises from the operation of a certain attractive power which has always the same limit; if, then, a stronger attractive power be presented, they will separate. The extremities of the Voltaic battery may be rendered respectively positive and negative to any extent by increasing the number of plates.—When two wires from the opposite extremities of such a battery are introduced into water, if their electrical states are more powerful than the natural electricities of its elements, these will necessarily separate, and pass to the oppositely electrified wires. The gases thus attracted to the wires will combine with them, if they are susceptible of combination; but if this is not the case, they will escape. Thus, when the wires are formed of a metal that readily combines with oxygen, no gas appears at the positive wire, but a quantity of the oxide of the metal is gradually formed there; and it has been observed, that when tellurium is employed for the negative metallic surface, a solid compound of that metal and hydrogen is formed.

The phenomena of Voltaic decomposition appear very simple, when considered in this way; for the reasoning applied in the case of water will apply in most other cases, as the bodies that usually appear at the positive wire are, for the most part, either compounds of oxygen, or of analogous properties, and may therefore be considered as naturally in the same electric state (negative): and the substances that appear at the negative wire are principally analogous to hydrogen, either from their actual inflammability, or from their containing a considerable portion of inflammable matter. Hence they may be considered as having the same natural electricity (positive).

These natural electric powers may indeed, in some instances, be exhibited. Touch with an insulated plate of metal some dry crystals of benzoic, oxalic, or other solid acid, and apply the plate with which the contact has been made to the insulated cap of a condensing electrometer; the leaves will open with positive electricity: hence it is fair to conclude the acid is negative, and this is agreeable to the hypothesis. Again, make a similar experiment with an insulated plate and dry lime, strontites, or barytes, and they will be found positively electrical. The same effect would probably be obtained by the alkalies, did not their rapid attraction for moisture interfere with the result.

It may also be observed, that those bodies which are capable of forming active Voltaic combinations, are, for the most part, such as are capable of combining chemically when their parts have freedom of motion: this is obvious in the arrangements of different metals; those which have the highest attraction for oxygen being positive with respect to all that have a less attraction for it; this is the case also with sulphur and the metals, and with acid and alkaline substances. Thus, in a combination of iron, copper, and an acid solution, the iron is the metal most affected by chemical action, and it is positive with respect to the copper; but in a combination of iron, copper, and an alkaline sulphuret, the copper is most affected, and it is then positive with respect to iron.

Substances that become electrical by contact, lose this power when combined. Copper and zinc, by mechanical touch, become electrical, but when fused together evince no electrical signs; and the same may be said of sulphur and copper, and of zinc and mercury.

An apparent illustration of this idea of natural electric energies may be obtained by an experiment originally contrived by professor

Lichtenberg, and since improved and explained by Mr. Cavallo and Mr. Bennett.

Experiment 9. Procure a resinous plate of 18 inches square and half an inch thick.* Draw the knob of a small Leyden bottle, charged with negative electricity, over one part of its surface, and the knob of a similar bottle, charged with positive electricity, over another part of its surface. Place the plate vertically, and project towards it from a spring powder puff, a mixture of red lead and flowers of sulphur. The mixed powder will be separated by the different electricities on the surface of the resinous plate. The red lead will adhere to the part touched by the negative bottle, and the flowers of sulphur to the part touched by the positive bottle. The figures they form are very curious, and always of different characters; they may be diversified in a very pleasing manner by describing letters or other figures with the knobs of the electrified bottles, or by com-

* It may be formed by melting together five pounds of resin, half a pound of bee's-wax, and two ounces of lamp-black, and pouring the mixture on a board having a rim round its edge to confine the composition whilst fluid. The blisters that form on the surface may be removed by frequently heating it before a fire, and suffering it to cool after each application of heat.

municating electricity to the resinous surface by conductors of any required form.

Experiment 10. This remarkable phenomenon of the separation of mixed powders by the action of the contrary electricities, can only arise from the actual electric state of those powders being different. This was first discovered by Mr. Cavallo, and may be thus exhibited.

Place a broad metallic plate on the cap of the gold leaf electrometer, and project some flowers of sulphur on it, either by an elastic-gum bottle, or spring puff, or even by shaking the sulphur through a linen bag; the electrometer will in a few moments open with negative electricity. Discharge the electrometer, remove the sulphur, and repeat the experiment with powder of red-lead, which should be made dry previously; the leaves of the electrometer will open with positive electricity.

This last result is stated on the authority of Mr. Bennett,* but it is a curious fact, that all the specimens of red lead I have hitherto tried, produce negative electricity when projected on the cap of the electrometer, though they are

* See his New Experiments in Electricity, p. 26; or the Philosophical Transactions for 1787, vol. lxxvii. p. 28.

attracted by the negatively electrified surface in Lichtenberg's Experiment. This anomaly can only be explained by supposing that the electricity of the red lead is different when it is projected with another powder. I state this circumstance, because the separation of the mixed powder of red lead and sulphur, or red lead and resin, has always taken place when I have projected them on a surface charged with both states of electricity; but either red lead, sulphur, or resin, separately sifted on the electrometer, has invariably occasioned it to diverge negatively.

The general accuracy of Mr. Bennett's experiments, and the coincidence of the greater number of them, with my own experience, lead me to believe, that the red lead he employed really produced the described effect; and there is probably a difference in that article resulting from various methods of manufacture. In Derbyshire, where Mr. Bennett resided, red lead is manufactured by the direct oxidation of the metal; but a considerable proportion of that sold in London is said to be made from Litharge, and is considered as less pure. This variety may account for the different results obtained by Mr. Bennett and myself, but it by no means explains the singular phenomenon of a negatively electrified powder being determined to a negative surface, when at the same distance from one that is positive!

The hypothesis of electric energy is supported by some other analogies. Thus the occasional evolution of heat and light, is common to both chemical and electrical action; and the developement of both chemical and electrical energy, is facilitated by elevation of temperature. But the most striking fact is, the power of promoting or suspending the usual operations of affinity by electric powers. Nitric acid, for instance, acts strongly upon copper; and according to the hypothesis, this arises from the copper being positive with regard to acids,* and experience shews, that by reducing this positive energy the action is really either lessened or suspended.

Experiment 11. Into a glass filled with dilute nitric acid, introduce a platina wire proceeding from the positive side of a Voltaic bat-

* It has been shewn, that the metals are positive and negative, with regard to each other, nearly in the order of their attraction for oxygen; but they are all positive with respect to acids, and negative with respect to alkalies.

tery. Connect a copper wire with the negative side of the battery; and complete the circuit by plunging the extremity of the copper wire in the nitric acid. There will be very little action, for the copper is rendered negative by its connexion with the battery; in proof of which, if it be separated from that connexion, it will be dissolved rapidly.

By a similar process, two substances that have no action on each other may be made to unite; there are many experiments of this kind; the following is one of the most simple.

Experiment 12. Fill a glass with a solution of sulphate of copper, and connect it with the positive end of a Voltaic battery. Immerse a slip of silver in the solution of copper, and suffer it to remain any length of time; no effect will be observed. Connect the silver with the negative extremity of the battery suffering it to remain in the solution, and in a few minutes it will be coated with copper.

In the same way various metals may be revived from their solutions, by others which have no natural attraction for them, until connected with the negative side of a Voltaic circuit.

The most material objection to the inference drawn from these experiments, appears to be the very slight electrical change that is adequate to the production of such phenomena, for they occur when a single pair of metals are associated together, and even when such an association consists of two slender pieces of wire, yet in such cases no electricity would be manifested even by the medium of the most delicate instruments.

Experiment 13. If a wire of silver and another of zinc, be immersed in a glass containing dilute muriatic acid, so as to remain at a little distance from each other, the zinc will give off hydrogen gas rapidly, but the silver will produce no effect. Bring the ends of the wires that are out of the acid in contact, by twisting them together; the quantity of hydrogen given off by the zinc will be diminished, and bubbles will be evolved from the silver.

If zinc, iron, or copper, are employed in the same way with gold, in dilute nitric acid, similar phenomena ensue, but the gas produced is nitrous gas.

Experiment 14. If a wire of iron and another of silver are immersed in a solution of copper, the iron will soon become coated with copper, but the silver will remain unchanged. Bring the wires in contact by twisting their upper extremities together, and the silver will be soon covered with a coat of copper.

Similar experiments may be made with a zinc and a silver wire, in solutions of lead, or tin.

Dr. Wollaston, to whom we are indebted for the two last experiments, has proposed the following explanation of them. " We know that when water is placed in the circuit of conductors of electricity, between the two extremities of a pile, if the power is sufficient to oxidate one of the wires of communication, the wire connected with the opposite extremity affords hydrogen gas.

" Since the extrication of hydrogen, in this instance, is seen to depend on electricity, it is probable, that in other instances, electricity may be also requisite for its conversion into gas. It would appear, therefore, that in the solution of a metal, electricity is evolved during the action of the acid upon it; and that the formation of hydrogen gas, even in that case, depends on a transition of electricity between the fluid and the metal.

"We see moreover, in the thirteenth experiment, that the zinc, without contact of any other metal, has the power of decomposing water; and we can have no reason to suppose that the contact of the silver produces any new power, but that it serves merely as a conductor of electricity, and thereby occasions the formation of hydrogen gas.

"In the fourteenth Experiment also, the iron by itself has the power of precipitating copper, by means, I presume, of electricity evolved during its solution; and here likewise the silver, by conducting that electricity, acquires the power of precipitating the copper in its metallic state."*

The experiments of this ingenious philosopher, by which the attraction of alkali, and the precipitation of copper on the surface of silver, were produced by the influence of negative electricity excited by the ordinary machine, have been already recited at page 191. They are considered by him as favouring the preceding explanation, and proving that oxidation is

* Phil. Trans. for 1801, vol. xci. p. 427, and following.

the primary cause of electric phenomena. To me they do not appear to favour any such supposition, but rather the contrary; for in the experiment with two different wires, touching each other, both produce the *same chemical effect*, yet, if they are electrical at all, the one is positive and the other negative, as all experiments on the association of different metals prove; and if two wires, that have no chemical action on the fluid in which they are immersed, be rendered respectively positive and negative, they are well known to produce *different chemical effects*.

But it is said the chemical effect produced by the silver wire, arises from electricity communicated to it by the zinc; and, that we have no reason to suppose that any new power is produced by the contact of the metals. Now, if this were the case, the mere conducting communication of the metals would be the only condition necessary to give the silver its chemical power; but the case is widely different; the communication must be not only *conducting*, but *metallic*, and even then no chemical effect will be produced, unless the extremities of the wires are immersed in the *same* liquid, or in two separate portions of liquid that have a *conducting communication* with each other.

Experiment 15. Place two glasses filled with a solution of copper near each other. Make a compound arc, by twisting together the end of a wire of zinc, with the end of a similar wire of silver. Connect the two glasses by placing the silver leg of the arc in one, and the zinc in the other. The zinc will immediately attract copper from the solution, but it does not communicate that power to the silver, though they are both closely connected. Whilst the compound arc remains, connect the two glasses by a second arc, formed of a piece of bent wire of any kind, except gold, or platina. The silver will be immediately covered with a coating of copper, and will continue to separate copper from the solution as long as the disposition of the apparatus remains the same. Now, the only difference in the arrangement, that appears to have operated as a condition to the chemical power of the silver, was the provision of another conducting communication between the glasses, in addition to that established by the compound arc; it therefore appears that the associated metals *cannot serve as conductors to the effect produced;* and indeed if they did, it would be scarcely possible any accumulation of power could result from the increased number of plates in a Voltaic battery

This experiment does not display any of the electric powers of a Voltaic combination; but it shews that the association of *three* different substances is essential to the chemical agency of such a combination; and the phenomena will be found to correspond with some experiments of Mr. De Luc, on the efficient groups in the Voltaic pile. This celebrated philosopher found that no chemical effects were produced by any Voltaic arrangement, unless two metals were employed with a liquid between them; and in the experiment last described, zinc, silver, and a metallic solution were *inactive, though in contact with each other, until the fluid was made the medium of conducting communication between the free extremities of the combined metals.*

The experiment last described will succeed, when the two glasses containing the metallic solution, are connected by any moistened conductor; but the chemical power of the silver wire will be evinced slower, in proportion as the length of the moistened conductor is increased; and in all experiments of the kind, the less the interval between the extremities of the compound arc, the more rapid is its action on the interposed fluid. Hence, in the arrangement of Voltaic apparatus, for the purpose of chemical decomposition, the ends of the conducting wires are placed at a greater or less distance from each other, in proportion as their action is required to be more or less intense.

Experiment 16. The arrangement of a simple Voltaic combination, by Mr. Sylvester, in which this effect is apparent, is represented by Fig. 41. It consists of a tall glass jar filled with very dilute muriatic acid. Through a cork placed in the neck of this jar two wires are inserted; the one a short straight wire of zinc, the other a long bent wire, of platina, or silver; by turning this last round, its upper end may be brought in contact with the zinc, or separated from it at pleasure When they are separate, the zinc only is acted on; but as soon as they are brought in contact, the platina or silver becomes covered with bubbles of gas, which appear soonest, and are evolved in the greatest quantity from the point S, and the part C; which are those separated by the least stratum of fluid from the zinc wire.

Notwithstanding this circumstance, the power of a simple Voltaic combination continues to exert its effect when the stratum of interposed fluid is considerable. If a tube of three feet long be filled with dilute muriatic acid, and a wire of platina be inserted through a cork in one of its extremities, and a wire of zinc in the other; on connecting the wires, gas will be soon evolved from the silver. If the tube be bent the effect will take place more slowly; but I have always found it occur. I took two similar tubes of eighteen inches long, and connected them by a short piece of flexible pipe, so as to form together a tube of three feet in length, with a joint in the middle, which admitted of its employment either as a straight tube, or as a syphon with a bend of any required inclination. In the open ends of this tube I placed respectively a zinc, and a platina wire; and found, that whenever their outer ends were connected by a wire, hydrogen was soon evolved from the platina; but this effect took place soonest when the tube was straight, and hence it appears that the power put in motion by these combinations, can pass more readily through any given column of a fluid in a straight line, than in any other direction.

It has been seen, that when any metal is in solution in the interposed fluid, it is revived by the wire which in other cases evolves hydrogen; and it has been shewn, by the effect of the silver and the platina wire, that metals which have no chemical action on the interposed fluid alone, may decompose it when combined with another metal. These facts, though far from being perfectly understood, may serve to explain some chemical effects which were before rather obscure. If a zinc wire, for instance, be immersed in a solution of lead, the latter metal will be revived in the form of a metallic vegetation, which increases gradually by additions to its extremities. The first separation of the lead is sufficiently intelligible; the acid in which that metal is dissolved, having a stronger attraction for the zinc, dissolves a portion of it, and deposits on its surface an equal portion of lead. But the lead, so revived, continues to revive more, and to receive additions at its remote extremities, whilst it would have been rather expected these additions would have been made on the zinc, and the vegetation that had been first formed protruded further into the fluid by that means. The contrary result is now understood to be obtained, by the revived particles of lead forming a Voltaic combination with the zinc and the surrounding fluid. This effect is analogous to that which obtains in various other instances.

Experiment 17. Spread a few drops of a solution of silver upon a pane of glass, and place a small piece of platina and a similar piece of copper wire upon it, at a little distance from each other. A vegetation will take place about the copper wire; but no effect will be produced by the platina. Bring the wires in contact with each other, and the Voltaic combination thus formed will occasion a beautiful vegetation of metallic silver to surround the platina wire.

With a solution of tin, and wires of zinc and platina, similar phenomena occur; but a considerable time elapses, after the contact, before the vegetation appears round the platina.

The immediate contact of the oxidable metal with the metallic solution is not absolutely necessary to the success of these experiments; it is only essential that a regular Voltaic circle, consisting of two different metals, and a moist conductor, be established.

Experiment 18. Fig. 42 represents a glass tube having a piece of bladder tied over its lower extremity water tight, and a cork inserted in its upper end with a platina wire passing through it. The tube is to be filled with acetate of lead, and placed in a small cup of zinc containing dilute muriatic acid; when a metallic communication is formed between this cup and the platina wire, the latter becomes studded with brilliant crystals of metallic lead. In this case the oxidable metal has no connexion with the metallic solution but through the medium of the platina wire on the one side and moist bladder on the other; but, on the principle of the 15th experiment, a somewhat similar result may be obtained when there is no connexion but through metal.

Experiment 19. Fill two similar glasses, the one with a solution of silver, the other with dilute muriatic acid; connect them by a compound wire arc of zinc and platina; the zinc being plunged in the muriatic acid, and the platina in the metallic solution. Immerse a second arc, formed of a bent silver wire, in the two glasses, one of its legs being in each; after some time the zinc wire will be entirely dissolved, and the platina will be found covered with minute crystals of metallic silver, displaying a very beautiful appearance under the microscope.

According to the hypothesis of electric energy, all the phenomena of decomposition and transfer are occasioned by the opposite electricities of the wires in the interrupted circuit, and the supposed natural electrical energies of the elements of all compound bodies. Of these energies more will be said hereafter; it is sufficient at present to observe, that the characteristic energies of oxygen and hydrogen have been entirely assumed from the phenomena of their separation, and appear to have been considered incapable of demonstration; yet, from the important and extensive action of these bodies, I should conceive such a demonstration is quite essential, before any reliance can be placed on the accuracy of the data on which the hypothesis is said to be founded. But with regard to the powers of the Voltaic apparatus, it may be asked, have we any evidence that the opposite electrical state of the wires in an interrupted circuit is essential to their chemical action? I believe, when all the phenomena are examined, not the slightest rational ground will be found for any such conclusion. As far as common electricity is concerned, it is obviously not so; for the strongest artificial electrization of the Voltaic apparatus has no effect on its chemical powers. Now it may be observed, that the electro-motive power of a Voltaic apparatus is too considerable to be overcome by the action of our electrical machines: but where is the proof of this? All the usual electrical effects disappear when the apparatus is electrified by communication with an electrical machine, their continuance can only therefore be inferred from the stability of the chemical effects, and such an inference would serve but as an argument in a circle.

Dr. Wollaston, indeed, succeeded in producing chemical changes by the action of an electrical machine, and with an arrangement nearly similar to that employed for Voltaic decompositions; different effects being produced by the wires connected with the opposite conductors. This shews a relation between the effects of the Voltaic battery and the electrical machine, but is no demonstration of the existence of electric energy; for the different electrical states of the wires can never be considerable, in consequence of the conducting nature of the fluids interposed between them; and it is consequently a far less probable cause of the phenomena they produce, than the current of electric fluid that passes from one wire to the other.

In reasoning on these phenomena, it should be always recollected, that no electrical effects of the Voltaic battery can be observed but when it is in an insulated state; that is, when its opposite extremities are unconnected by any conducting substance. Now the very converse of this obtains with regard to its chemical agency, which is never exerted but when a conducting connexion exists between the opposite ends of the battery; it is therefore, I think, more rational to conclude, that the phenomena arise from the circulation of some peculiar power, (which every experiment indicates,) than from an imaginary difference in the electrical state of the wires.

To demonstrate that the electrical state of the wires has no connexion with the chemical phenomena, Mr. De Luc contrived an apparatus in which a central wire was placed midway in water, between two wires proceeding from the opposite surfaces of a Voltaic apparatus; he had also a contrivance by which the actual electric state of the three wires could at any time be ascertained. When the end wires were respectively positive and negative the centre wire was neutral, yet the opposite extremities of this wire were *at the same time producing opposite chemical effects*; one end separating *oxygen*, and the other *hydrogen*. By a simple variation of the apparatus the central wire was rendered negative, and the negative end wire, neutral; yet they continued to produce the same chemical effects as before. Again, the central wire was rendered positive, and the positive end wire neutral; and still no change was observed in the chemical effects. The wire connected with the copper end of the battery continued to separate hydrogen, whether negative or neutral; the wire connected with the zinc extremity uniformly evolved oxygen, whether neutral or positive; and the centre wire separated *oxygen at one extremity, and hydrogen at the other, equally, when positive, negative, or neutral.**

* See De Luc's Analysis of the Galvanic Pile. Nicholson's Journal, vol. xxvi. p. 124.

I have made many similar experiments on an extensive scale, with batteries of from 100 to 1000 pairs of plates; which my attention to the proper means of exciting and employing such apparatus has enabled me to do with precision. The results I have obtained correspond very nearly with those recited by Mr. De Luc, and I cannot but consider his analysis, as by far the most correct and masterly investigation of the immediate phenomena of the Voltaic apparatus, that has been published since the original demonstration of its properties by Volta.

When a series of metallic wires are placed in a line at equal distances from each other, and are immersed in a fluid; on connecting the extreme wires with the opposite ends of a Voltaic battery, every wire produces a different chemical effect at each of its extremities; the ends that point towards the copper side of the battery separate oxygen, those that point towards the zinc extremity, separate hydrogen; and these opposite effects occur at each interruption of the metallic circuit, however numerous. Now it is scarcely possible, that a number of conducting wires, surrounded by a conducting fluid, can each have a different electricity at its opposite ends; and the obscure notion of an electrical polarity, (or induction,) which has been advanced to explain this anomaly, is quite incompetent; for no series of conductors can be made polar, or positive and negative at their opposite ends, but by the temporary derangement of their natural electricity, which can only obtain when they are separated from each other by some *nonconducting* substance; *and no one can maintain, that water, or any saline fluid, or acid mixture, is a nonconductor, either of the chemical, or electrical effects of the Voltaic apparatus;* yet the usual chemical changes produced by Voltaic electricity occur at every interruption of the metallic circuit in such fluids.

Experiment 20. Procure four glass tubes, one-fourth of an inch internal diameter, and four or six inches long, bent in the form of the letter V. Fill these tubes with blue cabbage liquor, and arrange them as represented by Fig. 43; the interrupted metallic circuit being formed through them by connecting arcs of platina wire. When the extremities of this apparatus are connected with the opposite wires of a Voltaic battery, after a short time the liquor in that leg of each syphon which inclines towards the copper extremity of the battery will become green, and that in the opposite leg red; and by reversing the connexions, those legs which were green, may be rendered red, and those that were red, converted to green.

That these phenomena depend on the transition of electricity from metal to water, and from water to metal, is (I think,) demonstrated by the following variation of the experiment.

Experiment 21. Remove all the platina wires but the two end ones, and connect the four syphons, (filled with blue test liquor as before,) by three arcs of moistened cotton. When the end wires have been sometime connected with the opposite ends of the battery, the liquor in the two syphons next the copper side, will be wholly changed to green, and that in the two syphons next the zinc extremity, will be wholly changed to red. Hence it is probable, that when the electric fluid passes from metal to water, it separates oxygen or acid; and when it passes from water to metal, it separates hydrogen, alkali, or inflammable matter.

The most difficult feature of all the Voltaic decompositions, is the invisible form, in which the separated elements of various compounds appear to traverse the fluid, and arrange themselves at the opposite wires. The oxygen and hydrogen that appear in some of our experiments at the distance of three feet from each other, are necessarily supposed to result from the same particle of water; and if this be situated at either wire, one of its elements, (either the oxygen or hydrogen,) must pass through the whole length of the tube to reach the other, and that in an invisible state, for the gases are separated at the opposite extremities without any apparent alteration of the interposed fluid. On the hypothesis of electric energy, the hydrogen is said to be attracted by the negative wire, because it is naturally positive; and the oxygen by the positive wire, because it is naturally negative; this does not explain how the same particle of water can have its elements liberated at so great a distance from each other; and to account for this fact, according to that hypothesis, it is necessary to suppose that the particles of water between the wires are arranged with their elements in juxta-position, like two parallel rows of beads, the one of hydrogen, the other of oxygen; as the decomposition goes on these are supposed to slide past each other, so that each particle of oxygen comes successively in contact with different particles of hydrogen. In other words, an atom of hydrogen escapes at the negative wire, and at the same moment an atom of oxygen is attracted to the positive; the number of atoms of oxygen and hydrogen between the wires are therefore still commensurate to each other, and have only changed their places.*

To me this supposition appears to increase the difficulty, for it infers a series of decompositions, and recompositions, of which we have no proof; and yet it does not seem probable that such phenomena could occur, without producing some apparent motion, or change in the interposed fluid. Besides, it cannot operate when the last particles of a saline compound are separated, and arrange themselves at the remote wires, or even in separate vessels; for, towards the conclusion of such an experiment, it is obvious no parallel rows of particles can exist.

Dr. Bostock has proposed an explanation, on

* See Dr. Henry on the Theories of Galvanic Electricity. Manchester Memoirs, vol. ii. New Series, p. 293; or Nicholson's Journal, vol. xxxv. p. 259.

the supposition that the separation of oxygen and other bodies at the positive wire, is occasioned by the union of the electric fluid with the other element of the compound, with which it forms an invisible combination and passes through the fluid to the negative wire; the electric fluid being strongly attracted by this wire, enters it, and deposits the hydrogen or other element with which it had previously combined; and this then becomes visible. This opinion is sufficiently ingenious, but it is liable to various objections.

1st. The attraction of the electric fluid for one of the elements of certain compound bodies in preference to their whole mass, and that too with a force equal to the subversion of their natural affinity, is an assumption perfectly gratuitous, and supported only by the phenomena it is advanced to explain.

2d. The invisible transmission of ponderable matter, as a consequence of its combination with the electric fluid, is purely hypothetical, and very difficult to conceive; more particularly when the substance so transmitted is a metal.*

* Dr. Bostock advanced this opinion some years ago, when the phenomena were less numerous; he applied it only to the

Notwithstanding these objections, I do not know that any more plausible explanation has been yet offered, for in many respects it is conformable to the observed phenomena. Voltaic decomposition never occurs but when a fluid forms the medium of connexion between the opposite wires; and almost all the experiments, as well as theory, indicate a current from one wire to the other.

Experiment 22. Procure a glass tube eighteen inches, or two feet in length, and half an inch diameter. Insert in it, by means of cubical pieces of cork, a series of wires, each an inch and a half, or two inches long, so that their ends may be about an inch distant from each other. Fill the tube with a solution of lead, and close its ends with two sound corks, with a wire

transmission of hydrogen. The statement above given differs therefore in some degree from his hypothesis; and bears equal resemblance in principle to one proposed at an early period by Mr. Cruickshank. See Nicholson's Journal, 4to. vol. iv. p. 257, &c. In the papers of this ingenious chemist, published in the above volume, it will be seen, that he developed the germe (if I may be allowed the expression) of the most important facts that have been since established relative to the chemical agency of the Voltaic apparatus.

passing through each. Introduce the tube into the circuit of a Voltaic battery, and in a short time that end of each wire, which points towards the negative side of the battery, will be covered with a vegetation of metallic lead; the direction of which appears to indicate the progress of some power through the tube, from the positive to the negative side of the battery. See Fig. 44.

Sir H. Davy has mentioned an experiment in which a vessel of water, containing a few globules of mercury, was made the medium of connection between the opposite ends of a Voltaic combination of 1000 plates weakly charged; the mercury was violently agitated, and a portion of oxide formed, which passed "in a rapid current from the positive towards the negative pole." No hydrogen was given off whilst the charge of the battery was moderate; but when the action was increased so as to evolve hydrogen, the globules of mercury became stationary; as if the same power that had given motion to the mercury was neutralized by, or employed in, the separation of the hydrogen.*

* Elements of Chemical Philosophy, p. 172.

Whatever be the true cause of the chemical phenomena of the Voltaic apparatus, its effects are invariable: at that wire of any combination, which in an insulated state affects an electrometer negatively, hydrogen, inflammable matter, or alkali, are sure to separate when the circuit is made through a fluid; and at the opposite wire, which in an insulated state affects the electrometer positively, oxygen or acid is as invariably found under similar circumstances. The regularity of these phenomena has occasioned a classification of chemical substances according to their electrical relations, which has been adopted by Sir H. Davy in his "Elements of Chemical Philosophy."

The indefatigable Berzelius, who appears to have been the first proposer of this arrangement, has denominated those substances that constantly separate at the negative wire, "Electro positive," and those that appear at the positive wire, "Electro negative," on the supposition that they are respectively in an opposite state of electricity to that of the wire by which they are separated. This nomenclature appears to me rather too hypothetical in the present state of our knowledge. We have no unequivocal demonstration of the existence of what have been called "natural electric energies," and considered as an "essential property of matter." The different electrical states, obtained by the contact and separation of different bodies, is certainly no evidence that they are naturally possessed of inherent electrical qualities. The operation is analogous to the usual process of excitation; and when two substances exhibit different electricities after such manipulation, the phenomena more probably result from the change of electrical capacity, induced by the contact of dissimilar bodies, than from any natural energies they possess. Besides, by an accurate performance of these experiments, I find the results are sometimes inimical to the hypothesis;* and it may also be observed, that we have many instances of pure chemical action in which no trace of electrical effect is ever discovered. But it is unnecessary to extend these objections; the hypothesis was mo-

* The result of my experiments on this subject will be given in a subsequent chapter.

destly proposed as a probability, and never has been obtrusively insisted on by its illustrious author; whilst the facts he has discovered, during its developement, are in themselves so truly valuable, as to demand the gratitude and admiration of every intellectual being.

CHAP. III.

Extensive Agency of the Voltaic Apparatus as an Instrument of Chemical Analysis. Its Influence in the Evolution of Light and the Production of Heat.

THE uniform action of the Voltaic battery in disuniting the elements of compound bodies, and determining different specific substances, invariably to the wires proceeding from its opposite extremities, offers a most advantageous and ready means of general analysis; which has been already applied with the happiest success, to the decomposition of an interesting class of chemical substances, and to the discovery of new and important agents.

The extensive experiments of Messrs. Hisinger and Berzelius,* confirmed by the researches of Sir H. Davy,† had demonstrated the constant separation of oxygen, and compounds in which it prevailed, at the wire pro-

* Annales de Chimie, tom. li. p. 172, &c.

† Phil. Trans. for 1807, p. 1, &c.

ceeding from the zinc surface; and of hydrogen and other inflammable matter, at that connected with the copper surface: at this latter, alkali was also frequently found, and from analogy it was in consequence concluded, that the alkalies probably contained a considerable proportion of some inflammable substance.

This conjecture was confirmed by Sir H. Davy in 1807: he found that a thin piece of potash, or soda, slightly moistened by exposure to the air, and placed between two conductors of platina, proceeding from the opposite sides of a powerful Voltaic apparatus, was resolved into a peculiar metallic substance highly inflammable, which appeared at the negative surface; and oxygen gas, which was evolved at the positive surface. By an extensive series of experiments, it was shewn that these bodies are in reality metallic oxides, and that the proportion of their constituent parts is somewhat different, being in round numbers, for potash six parts of metallic base to one part of oxygen, nearly; or it may be stated, that potash is composed of 86 parts of metal, and 14 of oxygen in each one hundred parts. The proportions in soda are nearly seven parts metal to two of oxygen; or 78 metal and 22 oxygen in each 100.*

The metal obtained from potash, is called Potassium; it is lighter than water in the proportion of eight to ten. At common temperatures it is solid, but soft and plastic. At a temperature of 150 it becomes fluid, and evaporates at a heat rather below redness. In colour it nearly resembles silver, but it tarnishes immediately when exposed in the open air, and can only be preserved under Naptha.† Its attraction for oxygen is so powerful, that it will detach that substance from almost all its combinations; and the result of this action is its consequent oxidation and reconversion into potash. If thrown upon water it immediately inflames, floats upon the surface, and burns with a mixed flame of white, red, and violet; rendering the water in

* For a full account of the experiments on the production of these metals, their properties, &c. see the very interesting paper in the Philosophical Transactions for 1808, p. 1, &c. or Nicholson's Journal, vol. xx. p, 290, &c.

† Naptha is a very light and sometimes colourless oil: it is found in a state nearly pure in some parts of Persia; but is usually obtained, for the purpose of experiment, by repeated distillation in a glass retort from a viscous substance called petroleum, which may be purchased at the druggists.

which the experiment is made alkaline. Similar phenomena ensue when it is brought in contact with ice. When moderately heated in oxygen gas it inflames and reproduces potash. Its action on water is always attended by the decomposition of that fluid; hydrogen is evolved, and the oxygen combines with the potassium to form potash. By measuring the quantity of hydrogen separated from water by the action of a given weight of potassium, the quantity of oxygen that metal combines with to form potash may be readily learnt. Each grain of potassium detaches about 1.06 cubic inch of hydrogen gas, and consequently combines with half that quantity of oxygen.

The metal obtained from soda is named Sodium; it is rather lighter than water, nearly as 0,9348 to 1000. It has the colour of silver; is less fusible than potassium, but tarnishes in air in the same way. It is fluid at the temperature of 200, and passes into vapour at a strong red heat. At common temperatures it is a soft metal, and a globule of it may be easily spread into a thin leaf by the action of a knife. It decomposes water violently, and floats on its surface, but does not inflame; the water is rendered alkaline, and when examined is found to contain pure soda. It acts nearly in the same manner as potassium, but with less energy on most substances, and must consequently be preserved under naptha. When thrown on the surface of nitric acid it inflames, and burns with great brilliance; it also occasionally scintillates when thrown upon hot water. The proportion of oxygen with which it combines to form soda, may be learnt by noting the quantity of hydrogen evolved from water by a given weight of the sodium.

Both these new metallic substances unite with mercury in various proportions, and form amalgams which decompose water, but more slowly than the metals themselves; these amalgams act upon all other metals, even platina and mercury.

The decomposition of the alkalies may, by care and attention, be effected with a battery of fifty pairs of plates of three or four inches square; but the results are rather uncertain. Two hundred plates form a very efficient arrangement; they should be excited by a weak acid mixture, (about one part strong muriatic, or nitrous acid to thirty parts of water.) A plate of silver or platina being connected with the negative side of the battery, a thin piece of pure potash or soda is to be placed upon it, and a platina or silver conductor proceeding from the positive side of the battery, is to be brought in contact with the upper surface of the alkali, which soon fuses at the points of contact: metallic globules shortly appear near the negative surface, and gradually increase in size, until a crust of alkali begins to form on their surface; at this moment they should be removed by the point of a knife, and instantly plunged under naptha; or if the experiment be merely intended to demonstrate their production, they may be brought in contact with the surface of water or nitric acid. It sometimes happens that no globules appear, but if the contact be preserved for some time, and the alkali be afterward raised, several will be found imbedded in its under surface. If the action of the battery be strong, it also sometimes happens that the globules inflame, and even detonate at the moment of their production; it is therefore adviseable not to bring the eyes too near during the experiment, or else to cover them with glasses. These experiments always require great care to insure their success, which a trifling variation in the power of the battery, purity of the potash, or moisture of the atmosphere, may prevent.—Soda is rather more difficult to decompose than potash, and therefore requires to be employed in thinner pieces; the pieces of potash should rarely exceed a quarter of an inch in thickness, and those of soda one-eighth of an inch.

To prevent the loss of the alkaline bases during their separation, by the powerful action of the air upon them, it has been proposed to effect the decomposition under naptha: the moist potash being placed between two plates of platina in a proper vessel, which is to be filled with naptha as soon as the contact with the battery is established;* in this way the action of the air is prevented, but the naptha decomposes, and hydrogen and charcoal are liberated, which renders the result less satisfactory than in the more simple form of the experiment. The most essential precautions are to preserve the alkali as dry as is consistent with a sufficient degree of conducting power, and to em-

* An ingenious apparatus for this purpose is described by Mr. Pepys, in the 31st volume of the Philosophical Magazine, page 241.

ploy the battery in a moderate state of action, in which it does not produce very intense heat, for that would destroy the metallic base at the moment of its production.

The amalgam of potassium, or sodium, with mercury, is easily procured, and may be obtained by a very moderate power. A glass tube, one fourth of an inch diameter and three inches long, having a short platina wire sealed in one end, is to have mercury poured into it until the end of the platina wire is covered; the rest of the tube is to be filled with a concentrated solution of alkali, either pure or carbonated. The platina wire, surrounded by mercury, is then to be connected with the negative end of a Voltaic battery, and the circuit completed by bringing a platina wire from the positive end, in contact with the solution of alkali Gas will be evolved from this wire, and the surface of the mercury will be greatly agitated; when the action grows weaker, the mercury may be poured into a glass of water, and the presence of the alkaline metal will be immediately indicated by the evolution of a cloud of minute bubbles of hydrogen gas, which may be collected by inverting over the mercury a small closed glass tube filled with water. This result I have frequently obtained with a battery of thirty pairs of plates of only two inches square.

The amalgam may be obtained more highly charged with the alkaline metal by employing a solid piece of alkali, with a small cavity on its surface, in which a globule of mercury is to be placed. The alkali is to be connected with the zinc surface of a battery, and the mercury with the copper surface; the mercury soon becomes more tenacious, and sometimes is converted into a soft solid, and in this state, if thrown into water, it produces a rapid decomposition.

The strong attraction of the metals of the alkalies for oxygen, renders them most active agents of chemical decomposition; by the strongest Voltaic power they can only be obtained in small quantity; and for the purpose of experiment they are now usually procured by another process first devised by the French chemists. A gun-barrel is bent nearly in the form of the letter S. An iron tube of the capacity of two cubic inches having a small hole at the lower extremity, and an iron stopper at the top, is ground into one end of the gun-barrel, and a tube of safety is fitted to the other. The iron tube is to be filled with pure dry potash, and the bent part of the gun-barrel nearest to it, with clean iron turnings: this part of the barrel is to be luted and placed in a small blast furnace; the iron tube projecting out on one side, and the vacant part of the gun-barrel, with its attached tube of safety, charged with clean oil, or naptha on the other. A strong heat is then to be raised in the furnace, and when the iron turnings have attained an intense white heat, a small furnace is to be applied to the tube containing the potash, which being fused thereby, will flow gradually through the small hole at the bottom of the tube, upon the iron turnings. The oxygen of the potash combines with the heated iron, and the potassium condenses in brilliant lamina in the vacant part of the gun-barrel, which must be kept cool by ice, during the process. As potash always contains water, that is also decomposed, and hydrogen escapes during the experiment, from the tube of safety; the cessation of this liberation of gas, is the sign for removing the small furnace from the tube, and the heat being raised in the blast furnace for a few minutes, as high as possible, to expel the last portions of potassium from the iron, the whole apparatus is suffered to cool. The gun-barrel is then to be cut at the commencement of the part which has been kept cool, for there the greatest portion of potassium is usually found; it must be detached by a chisel in as large pieces as possible, and introduced quickly into naptha, a portion of which fluid it is expedient to pour into the barrel as soon as it is first opened.

This process is attended with some difficulty, but it has been repeated successfully by many chemists in this country: a more detailed account of it may be consulted in the 32d volume of the Philosophical Magazine, pp. 89, and 276.

Another process, by the action of heated charcoal, has been employed by Curaudau; it is described in Nicholson's Journal, vol. xxiv. p. 37.

The composition of the fixed alkalies was entirely unknown before these experiments, but the volatile alkali, or ammonia, had been shewn to consist of hydrogen and nitrogen, in the proportion of three of hydrogen to one of nitrogen by volume. Now it is singular, that of three bodies whose properties are so analogous, two should be metallic oxides, and the third a compound of two gases; but there are experiments that seem to prove that either one or both of these gases contain a metallic substance, and that consequently ammonia may be, like the other alkalies, a metallic oxide!

Messrs. Berzelius and Pontin of Stockholm, discovered that when mercury is placed in a Voltaic circuit with solution of ammonia, the mercury being connected with the copper extremity of the battery, and the ammonia with the zinc, the mercury gradually expands to four or five times its original volume, and becomes a soft solid, nearly of the consistence of butter, having its metallic characters quite unimpaired. It is very remarkable, that by this change it gains only about one-twelve-thousandth part of its weight; yet has its specific gravity so much diminished, that from being thirteen or fourteen times heavier than water, it becomes only three times heavier. By a short exposure to the atmosphere it regains its original size and fluidity, absorbing oxygen, and reproducing ammonia. When thrown into water a similar effect is produced, the water being decomposed and hydrogen liberated.

These phenomena are very analogous to those observed with the fixed alkalies; some substance combines with the quicksilver and alters its properties materially, without impairing its metallic character; now, according to all existing analogies, this substance must be a metal, and this metal in returning to the state of alkali, absorbs oxygen, as is seen by its action on water. Hence it appears that ammonia consists of oxygen, and a peculiar metal, which may be called ammonium; but its analysis by other means evinces only the two gases, hydrogen, and nitrogen; the former of these being the lightest of all gravitating bodies, is most probably a simple or elementary substance; and on such a view, it would seem that nitrogen, though a gaseous body, is a compound of oxygen, and a metal.

The amalgam of ammonium, may be formed most readily by making a cavity in a moistened piece of muriate, or carbonate of ammonia, connected with the positive side of a Voltaic battery, and inserting in it a globule of mercury connected by a platina wire with the negative surface; in a few minutes a soft amalgam is formed; it must be transferred into water as quickly as possible when its action on that fluid is to be observed, as it changes by the shortest possible contact of the air.

Sir H. Davy has observed, that the strong attraction of potassium for oxygen, enables it to decompose ammonia even more rapidly than the Voltaic battery; and if an amalgam of potassium and mercury be placed in a cavity in moistened muriate of ammonia, it immediately increases in size, and becomes more consistent.

As some of the substances called earths resemble the alkalies in various properties, it was conjectured, that they also were metallic oxides; and this conjecture has been partly verified by the experiments of Messrs. Pontin and Berzelius, and Sir H. Davy. If a paste be formed with water, and either barytes, strontites, lime, or magnesia; and this paste be connected with the positive side of a Voltaic battery, and touched with an iron wire proceeding from the negative surface, the wire obtains the property of decomposing water.

If a globule of mercury be placed in a cavity in the earthy paste, and touched with a wire proceeding from the copper end of the battery, (the paste being connected with the zinc,) an amalgam will be soon formed, which has the property of decomposing water, and forming with it a solution of the earth employed. If this amalgam be introduced into a little tube made of green glass, and bent in the form of a retort, then filled with the vapour of naptha and hermetically sealed: on the application of heat to the end of the tube containing the amalgam, the mercury will distil over and leave the pure metal of the earth behind. This process is rather difficult, and requires great care; Sir H. Davy has by its means obtained an acquaintance with some of the properties of these metallic bases, but they have never been obtained in sufficient quantity to admit of a very accurate exa mination.

The amalgam from barytes, strontites, and lime, may be obtained with a battery of from 100 to 200 four-inch plates, in moderate time; that from magnesia requires a longer continuance of the action of the battery, and the other earths do not yield to its powers. These metals are named from the earths of which they appear to be the bases, as follows; namely, that from barytes, barium; strontites, strontium; lime, calcium; magnesia, magnesium; alumine, aluminum; silex, silicum, &c.

The decomposition of the alkalies and earths which had previously resisted very numerous attempts at analysis, are a monument of the importance of the Voltaic apparatus as an instrument of chemical research; and a proof of the ability with which it has been employed, which will be regarded with admiration and applause, as long as science shall continue to be cultivated.

The phenomena that have been described as the consequences of Voltaic decomposition obtain in every variety of experiment. Sulphuric acid introduced into the Voltaic circuit, gives off oxygen gas, and sulphur is deposited. Phosphoric acid evolves oxygen gas, and phosphorus combines with the negative wire. Ammonia separates into hydrogen and nitrogen with a small proportion of oxygen. Oils, alcohol, and ether, when acted on by a powerful battery deposit charcoal, and give off hydrogen, or carbonated hydrogen. And Mr. Brande has shewn, that when animal fluids containing albumen, are placed in the Voltaic circuit, the albumen is separated in combination with alkali at the negative wire, and in combination with acid at the positive wire. And, that with a powerful battery, it separates at the negative wire in the solid form; and with a less power, in the fluid form, so that it is probable animal secretion may depend on some such power.*

The effects that have been hitherto described result from the introduction of fluid bodies into the Voltaic circuit, and are nearly allied to the usual operations of chemical affinity. I have now to notice its action on solid conductors, inflammable substances, and gases.

When the opposite extremities of a powerful Voltaic apparatus are connected by a wire, at the moment of contact a distinct spark is perceived, which occurs every time the contact is alternately broken and renewed. If the contact is made with a wire terminated at the end, by a piece of well-burnt charcoal, the spark is considerably more vivid. And if two wires proceeding from the opposite ends of the battery are armed with charcoal points,† and brought

* Phil. Trans. for 1809. p. 385, &c.

† Charcoal for this purpose is usually made from box-wood, cut into pieces of about an inch long, and three-eighths of an inch thick. The pieces of wood are to be put into a crucible,

in contact with each other, the light evolved is more brilliant and intense than any that has been procured by other artificial arrangements. When the battery is powerful, the emission of light may be kept up for a considerable time; it is so dazzling as to fatigue the eye even by a temporary glance, and when it ceases, leaves the most brilliantly illuminated room in apparent darkness.

This light appears to be principally derived from the immediate action of the Voltaic apparatus, and not from the combustion of the charcoal; for, though that is partly ignited, it suffers comparatively but little waste, and the light is evolved with equal splendour when the experiment is made in gases which contain no oxygen; and will even take place, though with diminished energy, under water, alcohol, ether, oils, and other fluids whose conducting power is not too great.

The influence of the Voltaic spark on various gases may be ascertained by the apparatus described at page 83, and represented by fig. 9,

covered with dry sand; which is to be placed in a fire, and kept red hot for one hour. Or the wood may be charred by plunging it beneath the surface of red hot lead.

the wires within the globe being terminated by pointed pieces of charcoal, instead of balls. When a globe of this kind has been exhausted and filled with sulphuretted hydrogen, on taking the Voltaic spark in it, the sulphur is separated, and deposited on the interior of the globe, and produces, during its separation, a very beautiful appearance.

Some other compound gases are similarly affected; phosphorus separates from phosphuretted hydrogen, and arsenic from arsenuretted hydrogen.

With the most powerful Voltaic batteries the striking distance of the spark, or interval at which it passes from one conductor to another, is very inconsiderable. Mr. Children measured this effect by means of a micrometer, attached to two polished points of platina, which were inserted in a receiver containing very dry air. With 1250 pairs of plates the points were brought within one-fiftieth of an inch of each other before the spark took place.* With a large apparatus employed at the Royal Institution, which extends to 2000 pairs of four-inch plates, points of charcoal were brought within

* Phil. Trans. for 1809, p. 36.

a thirtieth or fortieth of an inch of each other before any light was evolved; but when the points of charcoal had beeome intensely ignited, a stream of light continued to play between them when they were gradually withdrawn even to the distance of near four inches. The stream of light was in the form of an arch, broad in the middle and tapering towards the charcoal points; it was accompanied by intense heat, and immediately ignited any substance introduced into it; fragments of diamond, and points of plumbago disappeared, and seemed to evaporate, even when the experiment was made in an exhausted receiver; though they did not appear to have been fused. Thick platina wire melted rapidly, and fell in large globules; the saphire, quartz, magnesia, and lime, were distinctly fused.*

In rarefied air, the discharge took place at a greater distance, and the beam of light was made to pass through an interval of six or seven inches.

These phenomena may be exhibited on a smaller scale by means of 100 pairs of plates of six inches square, an apparatus which is well

* Elements of Chemical Philosophy, p. 153.

suited for all experiments of fusion and ignition.

The arched form of the stream of light passing between two charcoal points, is often very perceptible when the distance of the points does not exceed half an inch.

From the low intensity of the most powerful Voltaic apparatus, but little attention to insulation is required in the transmission of its effects. The conductors employed for this purpose consist of copper wires passed through a short piece of glass tube, which serves as an insulator to hold them by. Such conductors are represented attached to the battery, and placed on a glass plate to inflame gunpowder, in fig. 37.

As the charcoal points usually become ignited when the battery has moderate power, almost any combustible substance may be inflamed, if placed between them. Oils, alcohol, ether, and naptha, are decomposed when the points are plunged into them, and inflamed when they are brought near each other upon the surface.

Some of the most pleasing effects of the Voltaic apparatus result from its action on metals; if these substances in thin leaves, are made the medium of communication between the opposite ends of a powerful battery, they inflame, and by continuing the contact may be made to burn with great brilliance. The best method of performing these experiments, is to suspend the metallic leaves to a bent wire proceeding from one extremity of the battery, and to bring in contact with them a broad metal plate connected with the opposite extremity; the brilliance of the effect may be increased by covering the plate with gilt foil. Gold leaf burns with a vivid white light tinged with blue, and produces a dark brown oxide. Silver leaf emits a brilliant emerald green light, and leaves an oxide of a dark grey colour. Copper produces a bluish white light attended by red sparks; its oxide is dark brown. Tin exhibits nearly similar phenomena, its oxide is of a lighter colour. Lead burns with a beautiful purple light; and zinc with a brilliant white light, inclining to blue, and fringed with red. For the distinct appearance of these colours it is essential to make the contacts with metal; for if charcoal be used, the brilliant white light it evolves absorbs the colours produced by the combustion of the metals.

If a fine iron wire be connected with one extremity of a powerful battery, and its end be brought to touch the surface of some quicksilver connected with the other extremity, a vivid combustion both of the wire and the quicksilver results, and a very brilliant effect is produced.

If a fine iron wire of moderate length be made the medium of connexion between the extremities of the battery, it becomes ignited, and may be fused into balls; or if a platina wire is employed, it may be kept at a red, or even white heat, for a considerable length of time; which seems to prove that some power is continually circulating through it; but however powerful the battery, wires are never dispersed by it, as they are by the action of a charged surface.

If the slender wire be inserted in any fluid, and then introduced into the Voltaic circuit, the fluid may be made to boil.

It has been lately noticed, that if any two wires of different thickness are taken, on either of which a certain battery can produce ignition, a greater length of the thickest wire will be ignited than of that which is thinner. This effect may probably arise from the cooling influence of the air, for the surface of the thin wire is most extensive in proportion to its quantity of metal; and that the surrounding medium has an influence on the degree of ignition may be proved by another experiment.

Experiment 23. Stretch a fine wire of platina, withinside a glass receiver placed upon an air-pump, so that the air surrounding the wire may be removed or restored at pleasure. Ignite the wire to a dull red heat, by connecting its opposite extremities with the wires from a Voltaic battery, of sufficient power for that purpose. Rarefy the air by the action of the pump; and as the rarefaction proceeds the ignition of the wire will become more vivid, until at length it obtains a glowing white heat. Admit air into the receiver and the wire will lose its intense heat, and appear more dull than at first. Rarefy the air again; the ignition will increase. Restore it to its original density, it will again diminish. These effects may be repeated many times, and will maintain the same proportion to each other, though they are less intense at each repetition.

I have ignited platina wire in various gases, without obtaining any remarkable result, with the exception of one experiment, in which a platina wire stretched in a receiver filled with hydrogen gas, was split into a number of minute fibres the moment the connection with the battery was made. The result appears to have been accidental, and has not been obtained a second time, in numerous repetitions of the experiment under similar circumstances.

The power of a Voltaic apparatus increases with the number of plates it contains, within certain limits, but the limit is different for the various effects it produces, and varies also with the manner of employing the apparatus.

The effects have been stated by Volta to be in the simple ratio of the numbers, but very limited series only, had been put together at the time this statement was made; and there appears to be a loss of power when very extensive arrangements are employed. The pure electrical effects, and the force of the shock, I have always found increase with the number, and I have employed an arrangement of 1500. The power of chemical decomposition, and transfer, also continues to increase with the number when the battery is excited by dilute acid; but if it be charged with river water, the power does not increase after four or five hundred plates. The powers of ignition have increased in exact proportion to the numbers within the limit of one hundred plates,* beyond that limit there appears to be a loss of power; for Sir H. Davy found that one hundred plates ignited three inches of platina wire one-seventieth of an inch diameter, and one thousand similar plates charged in the same way ignited only thirteen inches.† From the uniformity of the results I have obtained, and their correspondence with the experiments of Van Marum and Pfaff, on the continent; and Dr. Wilkinson, and Mr. Cuthbertson in this country; I am disposed to think the igniting power would be usually proportioned to the number of plates, if they could be always applied with the same effect; but when the series is extensive, there are various sources of dissipation, and it is rather

* See Nicholson's Journal, vol. xxix. p. 29, &c.

† Elements of Chemical Philosophy, p. 156.

difficult to render the large proportion of acid mixture then required, of uniform strength.

The French chemists have investigated the ratio of increase for different numbers of plates, as indicated by the quantity of gas liberated by the decomposition of water; and they announce that the increase is as the cube root of the number of plates.* The apparatus they employed, was arranged in the form of troughs of a particular construction, being part of a large apparatus constructed by order of the French government. Sir H. Davy states, that he has made similar experiments with the large combination of Porcelain troughs employed in the Royal Institution, and the results he obtained, indicate an increase nearly as the squares of the numbers.

The result of every experiment of the kind must be uncertain if a series of minute attentions are not observed, which appear to have been overlooked in those already instituted. The vessels employed for the decomposition should be of the same size and form; the wires of the same length and thickness, and placed at equal distances from each other, in a fluid of uniform conducting power.

* Recherches Physico-Chimiques, p. 30, &c. vol. i.

When the size of the plates is increased, their effects on perfect conductors, such as metals, charcoal, and strong acid solutions, are greatly augmented; but their action on imperfect conductors, as water, and various weak saline solutions, remains unaltered. If a battery, for instance, of thirty pairs of plates of two inches square, be compared with another battery of thirty plates of six inches square, charged with diluted acid of the same strength; there will be no material difference in the shock they produce, or the quantity of water they decompose in a given time; but the small battery will not melt wire, or burn metals, and will scarcely produce a spark between two points of charcoal; whilst the large battery will evolve a brilliant light between the charcoal points, deflagrate metallic leaves with rapidity, and ignite several inches of wire.

This remarkable fact, which appears to have been first noticed by the French chemists, is susceptible of some explanation, (on the supposition that the phenomena are electrical,) by reference to what has been said in other parts of this work on the subject of quantity and intensity. If a Leyden jar, for instance, having a square foot of coated surface, be applied to an electrical machine with another jar, whose coated surface is equal to four square feet; after a certain number of turns of the machine, they will both be charged, and to the same intensity, for they will equally affect an electrometer. But the large jar will contain four times the quantity of electricity that the small one does, and will fuse sixteen times the quantity of wire.

Now, suppose an imperfect conductor, capable of transmitting only such a quantity of electricity as is adequate to the charge of half a square foot; and it is obvious either of the jars before mentioned, would produce the same effect on such a substance; for they both contain more than it can transmit, and its conducting power, which remains the same in both cases, limits the effect that can be produced by either. It is consequently found, that if several different sized jars are charged to the same degree, the shock is nearly equally painful when received from either of them.

Mr. Cavendish has stated, that "iron wire conducts four hundred million times better than rain or distilled water; that is, the electricity meets with no more resistance in passing through a piece of iron wire 400,000,000 inches long, than through a column of water of the same diameter only one inch long. Sea water, or a solution of one part of salt in thirty of water, conducts 100 times, and a saturated solution of sea salt about 720 times better than rain water."* It is therefore probable, that the power excited by a Voltaic apparatus, with plates of two inches square, is in quantity equal or superior to the conducting capacity of most aqueous fluids, and consequently no increased effect can be produced on such fluids by larger plates, which increase the quantity of that power, but not its intensity. But if a conductor be presented to the large plates which is capable of receiving the increased quantity they furnish, the effect must necessarily be greater on such conductor in proportion to the increased impulse it may be supposed to receive. These facts are capable of easy illustration.

Experiment 24. Let two wires, proceeding from the extremities of a battery of fifty or one hundred plates of two inches square, be plunged in separate glasses of water, if the glasses are connected by putting a finger into

* Phil. Trans. vol. lxvi. p. 198.

each of them a shock will be felt at the moment of contact. Connect the water in the glasses by some fibres of moistened cotton, or by an inverted syphon filled with water; on repeating the contact with this arrangement, either no shock, or a very slight one will be felt. Make a similar experiment with another battery of the same extent, but with plates of six inches square. The shock will be nearly as great when the glasses are connected by moistened fibres, as when no connexion exists between them; and whilst the circuit exists through the moistened fibres, and the human body, if a second circuit be formed through a fine wire, several inches of it may be ignited. The imperfect conductors being incapable of conducting more than a small portion of the power excited by the large plates.

Whatever be the cause of the power of the Voltaic apparatus, I should conceive that the quantity of that power excited by any given number of plates under similar circumstances, will be in direct proportion to the size of the plates; and if the power be electricity. or should obey the same law that operates with charged surfaces, the comparative action of different sized batteries, containing the same number of plates, should be, with regard to their power of igniting wire, in the proportion of the square of the increased surface; thus if two batteries are taken, one containing fifty plates of twenty square inches surface, and the other fifty plates of forty square inches, the latter ought to ignite four times the length of wire, the former can ignite. From some experiments with plates of four inches square, and others, with plates of eight inches square, made many years since, it has been stated by Dr. Wilkinson, "that the power of ignition, in batteries of the same total surface, but with plates of different sizes, increases in the proportion of the squares of the surfaces of the elementary plates, singly taken in each."* It was afterwards shewn by Mr. Harrison of Kendal, that when the total surfaces are not equal, the rate of ignition must be as the sixth power of the diameters of the elementary plates, or as the cubes of their respective surfaces.† It appears also from some experiments with large plates, mentioned by Sir H. Davy, that the power of ignition for equal

* Nicholson's Journal, vol, vii. p. 207.

† Ibid. vol. ix. p. 242

numbers of plates probably increases in a higher proportion than the squares of their surfaces; for twenty double plates, each exposing a surface of eight square feet, ignited more than sixteen times as much wire as twenty double plates having each a surface of two square feet.*

Experiments of this kind should be made with batteries that have never been employed before, for the least difference in the state of the plates will have a material influence on the results obtained. Trials should also be made with various sized plates, increasing in regular progression from the smallest that are capable of igniting moderate-sized wire to some of at least a foot square.

Some inquiries of this kind have, I believe, been instituted by Mr. Children; who has arranged some gigantic batteries, having the largest plates hitherto constructed. The first arrangement consisted of twenty pairs of copper and zinc plates, each plate four feet high and two feet wide, placed in a wooden trough covered with cement: the quantity of fluid required to excite it was 120 gallons. The effect

* Chemical Philosophy, p. 156.

of this enormous battery on imperfect conductors was very inconsiderable, and it did not affect a gold leaf electrometer; but it produced powerful effects of ignition; rendered three feet of platina wire, 1-30th of an inch diameter, perfectly red hot, so as to be visible by strong day-light; melted completely eighteen inches of the same wire; ignited points of charcoal, and evolved a most brilliant light; but its agency in chemical decomposition was scarcely perceptible.*

More recently, the same active philosopher has put together a battery of similar extent, but with plates of two feet eight inches wide, and six feet long. The plates were fastened to a beam, which was suspended by counterpoises from the ceiling of his laboratory, so that the plates could be easily raised or let down into the cells. With this apparatus six feet of thick platina wire was ignited, and shorter pieces melted with facility. Iridium was melted into a globule, and the ore of iridium and osmium was partially fused; the heat produced appears to have been more considerable than has been hitherto obtained by any other means.

* Philosophical Transactions for 1809, p. 32.

A very powerful Voltaic apparatus has been recently constructed by order of the French government. It consists of 600 pairs of plates, each near eleven inches square, and consequently exposing together near 500 feet of surface; they are arranged nearly on the principle of the trough already described as the invention of Mr. Cruickshank.* Besides this extensive arrangement of large plates, there is a series of 1500, of a smaller size. No experiments on the igniting power of this battery appear to have been made; its action having hitherto been principally applied to imperfect conductors and to the production of electrometrical effects. The first experiments were on the comparative action of different fluids, both as exciters of the power of the battery, and mediums for the exercise of its chemical agency. The fact I have before adverted to, namely, that the electrical and chemical energies of a battery are never at their maximum together, but require different conditions for their production, was noticed during many of these experiments. Twenty-four of the large plates, excited by a mixture of acid and water, decomposed the alkalies, and

* Page 323.

effected many other chemical changes, distinctly; but produced only a slight effect on the electrometer by the aid of the condenser. The whole series of 600 large plates, when charged with water, did not produce any similar chemical effects, but their electrical power was much more considerable. The best conducting fluids were found to be most active as exciters of the chemical powers of the battery; thus saline solutions acted more powerfully than water, acid mixtures more powerfully than saline solutions, and strong acid mixtures were more powerful than those more diluted. In some instances, a mixture of acid and salt was more active than acid alone; and when acid mixtures of various strength were successively employed, the quantity of gas disengaged by two platina wires, from the same fluid, were nearly proportioned to the increasing strength of the acid used to excite the battery. The fluids which were most efficient in exciting the chemical power of the battery, were most rapidly decomposed when exposed to its action; and their excellence in this respect appeared to be proportioned to their conducting power.—Saline solutions were found to conduct better than water, alkalies better than salts, and acids better than alkalies. With sulphate of soda a curious result was observed; the facility of decomposition was as the cube root of the quantity of salt contained in the solution; so that if one solution of sulphate of soda yield twice as much gas as another in the same time, and with the same power, it is found to contain eight times as much salt. In the employment of different series of plates for the decomposition of water, the effect was found not to increase by any means in proportion to the number of the plates; so that when many batteries are applied to this purpose, it is better to employ them separately, on different portions of water, than collectively on one portion.

With batteries of different sized plates, charged with diluted sulphuric acid, the quantity of gas disengaged from weak nitric acid, by any given number of plates, was nearly in the proportion of their respective surfaces.

The large battery of 600 plates was usually excited by a mixture of 1 part sulphuric acid, 70 parts water, and 9 or 10 parts of common salt. The shock from it, when taken by an individual, was exceedingly painful, but when received by three or four joining hands, was much more moderate. The intensity of the battery in this state of action was such, that a spark could be obtained from water by its means; and the action was always too violent at first to permit the successful decomposition of the alkalies, which may be better accomplished with a lower power.

In twenty minutes the powers of chemical decomposition usually ceased; but the shock was still produced very powerfully, and the electrical effects continued without any diminution.*

From these experiments it appears, that the chemical action of a Voltaic battery is greatest when the fluid by which it is excited has most conducting power, and the greatest facility of decomposition; but that these circumstances have not the same influence on its electrical powers.

* Recherches Physico-Chimiques, vol. i. pp. 1 to 50.

CHAP. IV.

Sketch of the State of Theoretical Knowledge in Voltaic Electricity. Structure and Properties of the Electric Column.

In the preceding Chapters the most important properties of the Voltaic apparatus have been described, and in some cases the opinions that have been advanced in explanation were considered; it is therefore unnecessary to expatiate very extensively on this part of the subject, in which our knowledge is still very imperfect: a statement of the facts, which must be considered in every rational theory, promises therefore to be more useful than a detailed exposition of the various theories that have been proposed.

In all our Voltaic apparatus, there is a combination of three different substances in contact with each other, in successive groups; in general it is an arrangement of copper, zinc, and some conducting fluid. It is demonstrable, (as has been shewn,*) that the primary source of the electrical power of the apparatus, is the association of the two metals; and according to Volta, the interposed fluid serves only as a conductor of the effect of one pair of metals to another. As far as electricity is concerned, this opinion appears to be correct, for the electrometer is acted on, whatever be the nature of the interposed fluid,† and the degree of divergence is proportioned to the number of the plates. The electrometrical effects prove also, that the arrangement of a series of zinc and copper plates, with an interposed fluid, forms a conducting column, which in its insulated state is positive at one extremity, negative at the other, and neutral in the middle. This may be easily shewn by three gold leaf electrometers, connected at the same time with an apparatus of three or four hundred pairs of plates. The elec-

* See page 316 to 318.

† Thus Volta found the same effect on the electrometer when his apparatus was charged with pure water or with brine. And in the experiments of the French chemists, the large battery produced the same electrometrical effect when first charged with diluted acid, which had a strong chemical action on the plates, and when the acid had dissolved so much zinc as to form a saline solution which had no chemical action.

trometer connected with the copper extremity, will diverge with negative electricity, that connected with the zinc end, will separate to the same distance positively, and that connected with the centre plate of the series, will not be affected. But if either extremity of the battery be connected with the ground by means of a wire, the leaves of the electrometer connected with it will close; and those of the central electrometer will open with the same electricity, and to the same extent, whilst those of the opposite extremity will have their original divergence increased.

Hence it appears there is a real electro-motive property in the apparatus, by which the zinc end constantly tends to become positive, and the copper end negative; and it is also obvious, that the extent of this operation at either extremity, is increased by connecting the opposite end with the ground. This last experiment, by which the central plate may be rendered either positive, negative, or neutral at pleasure, proves also that the interposed fluid never acts as an insulator, for if it did so these changes could not possibly occur.

As the contact of either surface of the battery with the ground, exalts the electrical state of the opposite extremity, the same circumstance may be presumed to take place with every pair of associated metals, when their surfaces are in contact with a conducting fluid. Whilst the apparatus is insulated, the first zinc plate can only act on the electricity of its associate, the first copper plate; but the second zinc plate, through the conducting interposed fluid, can act on both these, besides its companion the copper, and may therefore become more highly positive; and it is easy to conceive that such a repetition of action would be attended with an increase of effect proportioned to the number of plates, and that the electrical tension of either end must be increased by connecting the other with the ground.

To ascertain if this principle really operated with a single combination; I took a pair of circular plates six inches diameter, very clean and even, one of zinc and the other of copper, and each provided with an insulating handle. When both plates were held by their insulating handles, and the zinc was successively applied to the flat surface of the copper, and after each contact made to touch the insulated plate of a condenser of six inches diameter; twenty contacts were required to communicate such a charge to the condenser, as would occasion the leaves of a very delicate electrometer to separate to a quarter of an inch. But when the copper plate, instead of being held by its insulating handle, was simply laid on the hand, or on any similar conducting body; *ten* successive contacts of the insulated zinc plate, communicated a charge to the condenser, which occasioned the gold leaves to separate to the distance of *more than half an inch!* On repeating these experiments, with the variation of touching the condenser with the copper plate, held by its insulating handle, and brought in contact with the zinc plate, first insulated, and then uninsulated, similar results were obtained, but with the contrary electrical state. Hence the similarity of action in a single pair of metals, and a combined series is sufficiently proved; and the preceding statement of the manner in which the electrical power is supposed to increase with the number of associated plates, is rendered highly probable.

So far the phenomena are sufficiently simple and consistent, for those described are not materially influenced by the nature of the interposed fluid, nor do they occur, but when the extremities of the apparatus are unconnected with each other, and consequently capable of maintaining the opposite electrical states. But the chemical effects, the shock, and the power of ignition, take place *only* when the extremities of the apparatus *are connected* by some *conductor*, and are also materially influenced by the nature of the interposed fluid. If these effects then, are produced by electricity, they can only result from its *circulation* in the apparatus; and as there is no reason to suppose that the electromotive power of the associated metals ceases when there is a conducting communication between their opposite surfaces; but rather that it is accelerated by such a circumstance; that very acceleration may be the cause of the phenomena, and the effects observed correspond very nearly with such an idea; for if it be admitted that the connexion of the opposite ends of the Voltaic battery by a conductor, occasions a current of electricity from the positive to the negative, that current must be more rapid, in proportion as the conductor is more perfect. Now it is found that the chemical effects are most considerable, and more promptly produced in fluids of the highest conducting power; thus the quantity of gas liberated in a given time from common water, is greater than from distilled water; saline fluids furnish more than common water; solutions of alkali more than saline fluids, and acids more than alkalies: and as the effects of a simple combination are influenced by the same causes as those that operate with a series; the fluids that are susceptible of the most rapid decomposition, are also most active in exciting the chemical effects of the battery, when employed as the connecting medium between its plates.

Acids are of all other fluid bodies, excepting metals, the most perfect conductors, and the chemical effects of the battery is more powerfully excited by them than by any other substances; it is possible their chemical action on the zinc may have some share in modifying the quantity of electric fluid, or the rapidity of its motion; but it is certain the effects are not in proportion to the chemical action: sulphuric acid, for instance, acts as powerfully on the zinc as nitric or muriatic acid, but it is not so active in producing the chemical agency of the battery: in like manner the alkalies which exert a very trifling action on the battery, excite its powers with greater energy than many saline fluids which are more efficient as chemical agents.

The ignition of wire, and of charcoal, in the Voltaic circuit, is conformable to this view; these substances are the most perfect conductors known, and when made the medium of communication between the opposite ends of a battery, must accelerate its electro-motive power to the greatest extent. The rapid circulation of electricity thus obtained, produces ignition, if the conductor be not too large in proportion to the quantity of electricity; but within this limit, the effect will be greatest with the thickest wire, because the acceleration will be more considerable in proportion to the facility of transmission. There is perhaps no other view on which the continued ignition of wire, and the increased action of large plates is so intelligible.

The cessation of chemical agency, and igniting power, as the chemical action of the acids or other menstrua declines, may arise from the total change which then occurs in the nature of those fluids; their conducting power is much diminished, and they may possibly by the change in their chemical properties, acquire some faculty of electro-motion subversive of the effect of the combined metals.

These observations are offered, as the most plausible inference from the observed facts that has yet occurred to me; and I shall be very highly pleased, if they are the means of eliciting some more satisfactory explanation of these phenomena.

It has been sometimes supposed that the fluid by which the different pairs of plates in a battery are separated from metallic contact with each other, does not act as a conductor; but, that "with regard to electricity of such low intensity, water is an insulating body." Did the action of the battery, or the phenomena of fluid decomposition depend on a circumstance of this kind, that action, and those phenomena, would be necessarily produced most perfectly by the most insulating fluids, but the very converse of this is observed; and so little does the electromotion of the Voltaic apparatus depend on any insulation between its plates, that all its phenomena are produced when the cells of a trough are filled so as to overflow, and the different plates are consequently connected over their top edges by a considerable stratum of fluid. Nay, the effects continue, though with diminished energy, when all the plates are connected together by *metal*, and consequently when every part of the apparatus is equally conducting. I cut a number of thin slips of sheet copper, and bent them into the form of the letter U, so as to form a series of simple springs. I then introduced both legs of one of these springs into each cell of a Voltaic battery, so that it pressed forcibly against the copper surface on one side the cell, and the zinc surface on the other. Having, in this manner, made a regular metallic communication between every pair of plates in a battery containing 50 of three inches square, I filled the cells with a diluted acid, and found, that notwithstanding the total absence of insulation, water was decomposed with great rapidity, a vivid spark produced by charcoal points, and gunpowder inflamed; and on applying the condenser, a charge was communicated which occasioned the gold leaves of the electrometer to strike the sides of the glass.

This phenomenon appears the more extraordinary at first view, because it is well known that if the plates are all connected together by a thin wire, the effect is almost totally destroyed; but in such case, the opposite copper and zinc surfaces of *each pair* of plates are made to communicate with each other, and consequently their electricities circulate individually, instead of being propelled forward from one cell to the other. But in the arrangement above described, the metallic springs are in contact with the zinc surface of one pair of plates, and the copper surface of another; there is consequently no communication between the opposite surfaces of any individual pair but what arises from the association of the copper and zinc; and as their mutual contact produces a motion of the electric fluid from the copper to the zinc, it cannot operate as a conductor in the contrary direction. The effect is therefore only diminished in proportion as the copper spring, by placing part of the zinc plate between two copper surfaces, diminishes its electro-motive energy. This experiment appears to me a satisfactory proof of the electro-motive power produced by the association of the metals, and of its tendency to produce a current of electricity from one extremity of the battery to the other, and consequently a circulation of electric fluid when the opposite extremities are connected: it also proves that the electromotive power is influenced by the nature of the substance interposed between the different pairs of metal, and thus accounts in some measure for the different effect produced by different fluids. This last circumstance is an interesting subject of inquiry; some instructive facts respecting it have been detailed by Professor Berzelius, in an account of an ingenious experiment made to prove that oxidation is not the cause of the electricity of the Voltaic apparatus. The following is extracted from his description. "I took twelve tubes of glass, half an inch diameter and three inches in height, and closed at one end. I half filled them with a strong solution of submuriate of lime (such as is obtained from the residue after the preparation of caustic ammonia), and above this fluid I poured diluted nitric acid, with the precaution not to mix the liquids. I arranged these tubes in succession, and then took copper wires, round one of the extremities of each of which I had melted zinc, in order to attach a knob of that metal to that end. I immersed the zinc-coated ends of each into one of the tubes to the bottom of the submuriate, and then bended the upper ends of the respective wires so as to immerse them in the middle of the acid of each nearest tube. This arrangement consequently formed a series in the order following: copper, zinc, submuriate of lime, nitric acid; copper, zinc, &c. It is evident that the chemical affinity which produces oxidation at the common temperature, was here at the surface of that part of the copper which was in contact with the nitric acid; and that if this oxidation had been the primary cause of the electricity of the apparatus, *the pole of copper*, in this construction, *ought to have possessed the same electricity* (namely, the positive) *as the zinc pole in the common pile.* Before the extremities of this small apparatus were connected, the copper continued to be constantly dissolved in the acid, which it turned blue, and the surface of the zinc remained metallic and without any perceptible change. And lastly, I combined the poles, by means of silver wires, passed into a tube filled with a solution of muriate of soda. But I was greatly surprised to find the effect *directly contrary* to what the theory, which considers oxidation as the cause of the electricity of the pile, had led me to expect. The solution of the copper instantly ceased, and the zinc became covered with a mass of white oxide, vegetating on all sides in the form of wool. The pole of the copper produced hydrogen gas as usual, and the zinc pole caused an abundant precipitate of muriate of silver. The electric state, therefore, produced in this case an affinity, which at the ordinary temperature of the atmosphere is inactive, and caused another very active affinity to cease, which was already in operation; *and this could be effected by no other cause than that of the electricity produced by contact*, which occasions the electric charge of the pile, and disposes the affinities which shall be put into activity.

"This little apparatus was very powerful, and disengaged so large a quantity of gas, as would not have been exceeded by one hundred pairs of plates. But what could be the cause of this?—I exchanged the submuriate for neutral muriate; it then produced a very moderate effect, corresponding with the number of pairs; and, lastly, I substituted neutral muriate of zinc instead of the muriate of lime, and then the effect was scarcely perceptible, though it continued sufficient to prevent the oxidation of the copper in the nitric acid, and to show that the conductor of the zinc pole continued always to be oxided."*

This experiment demonstrates the influence of the interposed fluid on the chemical effects of the apparatus, which may probably arise from its action on the electro-motive power produced by the association of the metals. It indicates also, that the chemical action of the battery is never exerted but when the electric fluid circulates from one extremity to the other; and corresponds in this respect with an experiment mentioned by Sir H. Davy, in which forty compound arcs of zinc and silver were arranged in the usual order, in a series of glasses, filled with a solution of muriate of ammonia rendered slightly acid by muriatic acid; whilst the extreme parts remained unconnected, no gas was disengaged from the silver, and the zinc was scarcely acted upon; but when they were connected, all the zinc wires were dissolved more rapidly, and hydrogen was disengaged from every silver wire.

* Memoirs of the Academy of Stockholm for 1812; or Nicholson's Journal, vol. xxxiv. p. 161.

In simple Voltaic combinations, it appears essential to the production of chemical effects, that there be a transition of the elements of the interposed fluid; and as this may be presumed to take place also in each cell of a battery, it is perhaps one cause of the superior action of those fluids which are most readily susceptible of decomposition. When, for instance, (as in experiment 19,) a compound arc of zinc and platina is placed with the platina leg in a solution of silver, and the zinc leg in dilute muriatic acid, no precipitation of silver takes place unless the glasses are connected by some fluid medium, or by a metal which is soluble in the acid of the solution of silver. With arcs of platina, or gold therefore, no effect is produced, either in this arrangement, or that of the fifteenth experiment; but with any other metal, a portion of the silver or copper of the solution is revived, and a corresponding portion of the simple connecting arc is dissolved, and occupies the place of the revived metal in the solution. Hence the corrosion of the zinc plates in the Voltaic battery, and the liberation of hydrogen at the copper surfaces.

From the phenomena hitherto described, it

appears that the primary source of the electric power of the Voltaic apparatus may be considered to be the association of the metals of which it is composed; but the chemical effects, though probably arising from the same cause, are obviously influenced by the nature and action of the interposed fluid. The relation of the various parts of a Voltaic apparatus, (as usually constructed,) to the various effects it produces, have been recently developed by the masterly experiments of Mr. De Luc.* The ordinary apparatus consists of three constituent parts, namely, two metals and a fluid, being usually when arranged in a pile; copper or silver, zinc, and wet cloth, following each other in successive groups. Now, if these be regarded attentively, without any regard to Volta's theory, they may be considered as divided into ternary groups under three different aspects.—1. Zinc and silver with wet cloth *between* them.—2. Zinc and silver *in mutual contact* with wet cloth on the side of the zinc.—3. Zinc and silver still in mutual contact, but the wet cloth on the side of the silver. Either of these ternary associations may be the cause of the action of the apparatus;

* Nicholson's Journal, vol. xxvi. p. 113, &c.

but the really efficient groups may be ascertained, if each of the ternary associations are successively mounted as a pile, the different groups being separated from each other by some conductor that does not materially affect their electro-motive power. Mr. De Luc employed for this purpose small tripods, formed of brass wire so bent, as to touch the plates between which the tripod was placed, only at the three points of support.

The first dissection of the pile by this method was to form an arrangement of seventy-six groups of zinc and silver with wetted cloth between them; one group being placed first, (suppose with the zinc plate lowest,) then upon the silver plate a tripod of brass wire; upon that another group with the zinc plate lowest; again, upon its silver, a tripod, upon that a third group in the same order, and so on until the whole seventy-six groups were arranged.

Under these circumstances the same chemical and electrical effects were obtained, as when the apparatus was put together without the brass tripods. It therefore appeared that the efficient groups for all the effects of the apparatus, were an association of silver and zinc, with wetted cloth between them. To ascertain the truth of this indication, a second dissection was made. In this the two metals were placed in contact with each other, and the wet cloth in contact only with the zinc plate. Suppose a pair of zinc and silver plates in contact with each other, placed on the base of the pile with the silver lowest, then a disk of wetted cloth upon the zinc, and a tripod upon the wetted cloth; then another group of zinc and silver, with wet cloth upon the zinc; then again a tripod, and so on in regular order, until the seventy-six groups were arranged.

With this apparatus the electrical effects were produced as before; but though these ceased when the usual glass tube for decomposing water was made to connect the opposite poles, not the slightest chemical effect was produced.

From this it appears, that the condition for the production of chemical and electrical effects is different, the latter requiring the arrangement of silver and zinc in mutual contact, the successive pairs being separated by a moist conductor, which may be in actual contact with the zinc only; the former requiring the association of silver and zinc, with wetted cloth *between them.*

A third dissection of the pile was thus arranged: silver and zinc in mutual contact, wetted cloth in contact with silver: 76 of these groups were placed in regular order, with a tripod upon the wet cloth of each group, as in the former experiment. With this arrangement neither chemical nor electrical effects were produced; the absence of electrical signs Mr. de Luc ascribes to the zinc plates being in contact on one side with the silver, and on the other with the brass of the tripod, which he regards as a counteracting effect. The absence of chemical signs arises from the want of the condition for their production, namely, successive associations of zinc and silver, with a fluid between them and in contact with both.

When either the continuous pile or that composed of the efficient ternary groups, are put together with the pieces of cloth moistened with pure water, although chemical effects are produced, no perceptible shock can be felt; but when the pieces of cloth are moistened by a solution of common salt, the shock is very distinct. Hence Mr. de Luc concludes, that for the production of chemical effects in the circuit it is essential that the zinc undergo oxidation, and for the production of the shock it is necessary that such oxidation be effected by the action of an acid.

Mr. de Luc conceives the phenomena of the pile may arise from some modification of the electric fluid which pervades it during the oxidation of the zinc; and as, in his experiments, he obtained more perceptible electrical indications by the aid of the condenser, from wires immersed in water, when the chemical effects and the shock were produced, he concluded that this modification of the electric fluid was attended by a retardation of its course, by which a very small quantity was enabled to produce effects which are not obtained by a much larger quantity when set in motion by the electrical machine.

This idea, it may be observed, is the very converse of that which, from a more general and extended view of the phenomena of the Voltaic apparatus, I have ventured to propose. It was indeed a natural inference at first view, from the experiment in question, when that alone was considered; but the increased rapidity of decomposition which always attends the increased operation of that influence, which is here supposed to cause a retardation of the current that occasions decomposition, is very inimical to any such supposition; and the usual phenomena of electrical analysis are equally at variance with it.

When any fluid is decomposed by the action of the common electrical apparatus, the effect is always proportioned to the intensity of the current of electricity that passes through it; and in the decomposition of water, when the metallic surface in contact with it is of moderate extent, very strong shocks in rapid succession are required. It is to the acute intelligence of Dr. Wollaston we are indebted for the means of executing this analysis with a more moderate power. He enclosed the metallic conductor in glass, or wax, and exposed only a very small portion of its surface to the fluid. The current of electricity being thus reduced in volume, was proportionably increased in force; and by rendering the exposed surface very minute, a sufficient intensity was produced, by a moderate quantity of electricity.

When a circuit is made through water, by

wires proceeding from the opposite extremities of a Voltaic battery, those wires can impart no charge to the condenser, unless the quantity of electricity evolved by the battery is greater than the water can transmit: therefore any cause that increases the quantity, will produce an augmentation of effect by this test, whilst the column of water remains the same: or if the velocity of the electro-motion of the apparatus be increased, whilst the same imperfect conductor is interposed between its extremities, a similar effect must take place; for the positive wire will receive electricity from the pile, faster than it can transmit it to the water, and the negative wire yields electricity to the pile, more rapidly than it can receive it from the water; so that a slight positive, and negative charge will be given to the condenser by these wires respectively, whenever the electro-motion of the pile supplies electricity faster than the water can conduct it; and the charge will be highest when the supply is most rapid. Now, according to the principle I have proposed, the most rapid electrio-motion of the apparatus, will be produced when the different pairs of plates communicate with each other through the medium of the best conducting fluids: it is therefore obvious, that the result of Mr. De Luc's experiment, in which a more considerable charge was communicated to the condenser by wires immersed in water, when the pile was excited by a saline solution, than when it was excited by pure water, is conformable to the principle I have stated; and the legitimacy of this inference is confirmed by a variation of the experiment; for if, when the apparatus is excited by a saline fluid, the tube that connects its extremities be filled with the same fluid instead of pure water, no increased charge will be given to the condenser by either of the wires, because the increased electro-motion of the apparatus is then compensated by the increased conducting power of the fluid by which its extremities are connected.

When different degrees of chemical action are excited in the Voltaic apparatus by the introduction of various fluids, I have always found that more powerful the action that is produced, the more transient is its duration. This circumstance is of importance in the practical application of the instrument, since it offers the means of judiciously applying various methods of experiment, and of continuing the action of the apparatus during any required time. When the battery is charged with water, its chemical action is feeble, but it appears to continue without diminution for an indefinite length of time; by the addition of a minute quantity of muriatic acid, $\frac{1}{500}$th part for instance, its chemical action is greatly augmented, and still continues for a considerable period. When the proportion of acid is increased to a 30th or 20th part, the action is considerable, but comparatively of short duration. I have found no solutions so advantageous as those of acids, and I prefer the muriatic acid to all others; the nitric is indeed rather more powerful in the same proportion, but its cost is four times as great, and I have found that it destroys the copper plates as well as the zinc. The nitrous gas evolved by its action is also much more offensive than hydrogen, which results from the employment of muriatic acid.

The experiments of Mr. De Luc induced him to conclude with Volta, that the electrical effects of the apparatus result entirely from the successive association of the different metals, separated into pairs by some conducting substance that does not interfere with their electromotive power. To ascertain if a liquid was essential to this effect, he mounted a pile with pieces of cloth not moistened, and he found the electric effects were still produced, but somewhat weaker than with the wetted cloth. He then instituted a series of experiments, successively mounting the pile with different animal and vegetable substances, interposed between the pairs of metal, instead of wetted cloth. Of the various substances tried, he preferred writing paper, as the most convenient of those that were efficient. The apparatus constructed in this way was found to have the same electrical indications as the common Voltaic pile, but it produced no chemical effects, however numerous the pairs of plates; nor was any oxidation of the zinc produced by its most protracted action. These circumstances led to the idea, that by the extension of the number of groups, a kind of perpetual electric machine might be formed; and, as in the previous trials, it had been found that the effect was rather increased by pasting the paper upon the silver or copper. Dutch gilt paper, which consists of thin copper leaf, laid upon paper, was employed instead of the usual silver, or copper plates, and moist conductors. Eight hundred plates of tinned iron being put together with the same number of Dutch gilt paper between them, the copper sides being all turned in one direction: the combination was found to affect the electrometer more powerfully than any Voltaic battery had been ever observed to do; but on the application of the usual glass tube with water, no chemical effect was noticed. The apparatus was left for a considerable time, and its action on the electrometer continued without diminution; and subsequent experience has shewn that it does so for any period during which the experiment has been continued.

Thus was invented a new and important Voltaic arrangement, highly valuable both in a theoretical and practical view: in the former, as separating the pure electrical effects of the Voltaic battery from its chemical power, and demonstrating the permanence of its electromotive faculty: in the latter, as providing a spontaneous and permanent electrical machine, in which the opposite electrical states perpetually exist, without any new excitement. Besides these properties, the new apparatus promises to become an important meteorological instrument; for the degrees of its electrical indications have been observed to vary with the different seasons of the year, and are probably influenced by some of the causes by which our atmospherical phenomena are produced.

To distinguish this instrument from the usual Voltaic apparatus, from which it differs in many respects, Mr. De Luc has proposed to call it "the Electric Cólumn," an appellation sufficiently appropriate, since the effects it produces are purely electrical.

I have made very numerous experiments on the construction of such columns, and have varied their combinations most extensively. The materials I prefer, are thin plates of flatted zinc alternated with writing or smooth cartridge paper, and silver leaf. The silver leaf is first laid on paper, so as to form silvered paper, which is afterwards cut into small round plates by means of a hollow punch. In the same way an equal number of plates are cut from thin flatted zinc, and from common writing or cartridge paper. These plates are then arranged in the order of zinc, paper, silvered paper with the silvered side upwards; zinc upon this silver, then paper, and again silvered paper with the silvered side upwards; and so on, the silver being in contact with zinc throughout, and each pair of zinc and silver plates separated by *two* discs of paper from the next pair. An extensive arrangement of this kind may be placed between three thin glass rods, covered with sealing-wax, and secured in a triangle, by being cemented at each end into three equi-distant holes in a round piece of wood; or the plates may be introduced into a glass tube previously well dried, and having its ends covered with sealing-wax, and capped with brass; one of the brass caps may be cemented on, before the plates are introduced into the tube, and the other afterwards; each cap should have a screw pass through its centre, which terminates in a hook outside This screw serves to press the plates closer together, and to secure a perfect metallic contact with the extremities of the column. The instrument constructed in this way is shewn by fig. 45.

Soon after the invention of the column, Mr. B. M. Forster discovered that, when a sufficiently extensive series was put together, its electric power was sufficient to produce a sort of chime by the motion of a small brass ball between two bells, insulated, and connected with the opposite extremities of the column. He constructed a series of 1500 groups, and by its agency kept a little bell-ringing apparatus in constant activity for a considerable length of time. If the electro-motive power of the apparatus be really permanent, as it appears to be, there is no doubt a perpetual motion may be thus produced. I have contrived an arrangement which is well calculated to ascertain this point, by excluding to a very considerable extent, the operation of extraneous causes of interruption, and it at the same time renders the disposition of the apparatus rather elegant. A series of from 12 to 1600 groups are arranged in two columns of equal length, which are separately insulated in a vertical position by glass pillars constructed on my new principle of insulation; the positive end of one column is placed lowest, and the negative end of the other, and their upper extremities being connected by a wire, they may be considered as one continuous column. A small bell is situated between each extremity of the column, and its insulating support and a brass ball is suspended by a thin thread of raw silk, so as to hang midway between the bells, and at a very small distance from each of them. For this purpose the bells are connected, during the adjustment of the pendulum, by a wire, that their attraction may not interfere with it; and when this wire is removed, the motion of the pendulum commences. The whole apparatus is placed upon a circular mahogany base, in which a groove is turned to receive the lower edge of a glass shade with which the whole is covered.—See Fig. 46.

I have an apparatus of this kind, containing only 1200 series, which was constructed fourteen months since, and has never ceased to ring, but when removed from one situation to another, which convenience has rendered necessary several times during that period. There was, however, one interval of more than six months in which it was never disturbed, and during that time its motion never ceased. Mr. de Luc has a pendulum which has constantly vibrated between two balls for more than two years, and its motion still continues.

If a column of about 1000 series is placed horizontally, with each of its extremities resting on a gold leaf electrometer, as shewn in

Fig. 47, the electrometers will each diverge; that connected with the zinc extremity of the column will be positive, that connected with the silver extremity will be negative. If the column be very powerful, the gold leaves of the electrometers will alternately strike the sides of the glass, but this motion is soon stopped by their sticking to it. If the simple divergence only is produced, on touching either extremity of the column, the electrometer connected with it closes, and that at the opposite extremity has its divergence increased. This is analogous to the effect of the Voltaic battery when disposed in a similar manner; but the motion in the column is slower, which (I suppose) arises from the more imperfect conductors of which it is composed.

There is some cause, not yet perfectly developed, that appears to influence the power of the column to produce the motion of light metallic pendula. In the bell-ringing apparatus, for instance, though the motion always continues, it is much more rapid at some times than at others, and the oscillation of the pendulum, though usually as uniform as that produced by mechanism, is on some occasions singularly wild and irregular. The frequency with which the gold leaves of an electrometer strike the sides of the glass, when connected with an electric column, is also different at different times: the variations observed in some experiments of Mr. de Luc are much more considerable than I have yet noticed, with the more powerful columns of my construction.

Mr. De Luc has proposed, as an interesting object of enquiry, to make regular observations on the action of the column, and the number of oscillations it produces in a given time, at each observation. For this purpose a single column of from one to two thousand series may be supported vertically on an insulating pillar. A bent wire with a ball at its lower end, is to be connected with the upper extremity of the column, so as to hang parallel with, and at some distance from it; the ball at its lower extremity being diametrically opposite to a similar ball that is screwed into the lower cap of the column. To the same cap there is also screwed a brass fork with a fine silver wire stretched between its extremities; this is placed above the ball and projects farther from the column, so that when the pendulum moves towards the ball it strikes this wire first, and receives a kind of jerk, which prevents it from sticking. The pendulum consists of a gilt pith ball suspended by a very fine silver wire, which hangs parallel to the bent brass wire, to which it is fastened at top; the arrangement is such, that the gilt pith ball would be always in contact with the brass ball that proceeds from the upper extremity of the column, if the apparatus had no electrical power; it therefore always returns to this situation, when, after being attracted to the lower extremity of the column, it discharges its electricity by striking against the cross silver wire. This apparatus (which Mr. De Luc has called, "aerial Electroscope,") is represented, covered with a glass receiver, by fig. 48.

There appears every reason to believe, that the action of a well constructed column will be permanent; I have several that have been constructed nearly three years, and they are still as active as at first. There is however a precaution necessary to their constant and immediate action; the two ends of a column should never be connected by a conducting substance for any length of time; for, if after such continued communication, it be applied to an electrometer, it will scarcely affect it for some time. It is therefore necessary, when a column is laid by, that it be placed upon two sticks of sealing-wax so as to keep its brass caps at the distance of about half an inch from the table or other conducting surface on which it is laid. And if a column which appears to have lost its action by laying by, be insulated in this way for a few days, it will usually recover its full power.

There is another cause of deterioration which is more fatal; it is the presence of too much moisture. If the paper be perfectly dry it is a non-conductor, and will not therefore produce any action in the column; but this perfect dryness can only be obtained by exposing the paper to a heat nearly sufficient to scorch it, and in its dryest natural state I have always found the paper sufficiently a conductor, even when, by exposing the paper discs to the heat of the sun, they have been so dried as to warp considerably. When the paper is sufficiently dry, the action of the column continues without diminution; and on taking such an apparatus to pieces after it had been constructed thirty months, no trace of oxidation was evident on the zinc plates.

I have formed some columns of very extraordinary power by various novel methods of combination; and have noticed some very singular phenomena during various experiments on this subject, in which I am still engaged; but the results are not yet sufficiently mature for publication.

The size of the plates in the column need not be large; I have constructed them of various sizes, and find no proportionate advantage by extending the diameter beyond 5-8ths of an inch; they may even be constructed much smaller than this.

By connecting the extremities of a column of at least 1000 series, with the opposite coatings of a Leyden jar, during a period of from one to five minutes, a charge is usually communicated to it capable of affording a small but distinct spark, when the discharge is made by a wire that is not very thick.

The most extensive series I have yet made experiments with, consisted of twenty thousand groups of silver, zinc, and double discs of writing-paper. Its power was considerable. Pith ball electrometers, with balls of one-fifth of an inch diameter, and threads of four inches long diverged to the distance of two inches and upwards, when connected with its opposite extremities. An electrometer at the center was not affected. When either extremity of the column was connected with the ground, the electrometer attached to that extremity closed, and the central electrometer opened with the same electricity, whilst that connected with the opposite extremity had its original divergence considerably increased; but the electro-motion was so slow, that some minutes were required to produce the full effect.

By connecting one extremity of the series with a fine iron wire, and bringing the end of this near the other extremity, a slight layer of varnish being interposed, a series of minute bright sparks were obtained by drawing the point of the iron wire lightly over the varnished surface.

A jar containing fifty square inches of coated surface was charged by ten minutes contact with the column, so as to convey a disagreeable shock, felt distinctly in the elbows and shoulders, and by some individuals across the breast.

The charge from this jar could perforate thick drawing-paper, but not a card. It had just power to fuse one inch of platina wire, of the five thousandth of an inch diameter.

Notwithstanding the considerable electric power of this combination, it had not the slightest chemical action; neither the best nor worst conducting media were affected. Saline compounds tinged with the most delicate vegetable colours, were exposed under the most favourable circumstances to its action, and in some instances for many days, but no chemical effect was produced.

It therefore appears indispensably necessary to the chemical power of the Voltaic apparatus, that a *liquid* be interposed between each pair of its plates, whilst for the pure electrical effects, the only condition appears to be the association of the two metals; and the connexion of the different pairs, by some conductor that does not interfere with their electro-motive power.

I am now constructing, and have nearly completed, an addition to the above series of columns, which will form an arrangement together of 60,000 groups; it was not possible for me to make the experiments with them in time for this publication; but my principal intention is to ascertain if any chemical effect can be produced by the most powerful column; if it cannot, I think the assigned condition for chemical action must be considered as established: and the determination of that circumstance will be one step towards a correct theory of Voltaic electricity.

The discoveries of Franklin displayed the influence of electricity in the production of the most magnificent phenomena of nature. That of Volta has led to the rapid developement of its connexion with her more silent, but important processes. Like the power of gravitation, it seems to apply more extensively, the farther its investigation is pursued. Like that power too, its nature may for ever escape our cognizance; but the contemplation of its effects may supply new facts calculated to extend the resources of art, and enlighten our conception of the infinite variety, and harmony, of natural phenomena. Such pursuits are amongst the best sources of intellectual improvement, for they call into action the highest powers of the mind, and present a constant succession of interesting objects for their exercise.

APPENDIX.

Various Additions and Corrections. Experiments on the Electrical Effects evinced after the Contact of Dissimilar Bodies.

1. Sources of Electrical Excitation.

An useful supplement to the table on the effects of friction between various bodies, (given at page 33,) will be found in the experiments of Mr. De Luc on the same subject, which are detailed in the 28th volume of Nicholson's Journal, p. 1. He obtained a determination of the electrical effects produced by friction, by constructing a minute machine, in which various bodies were applied to each other as cylinder and rubber; the rubber being insulated, and an insulated conductor receiving the electricity of the cylinder, so that the effect produced on both could be correctly ascertained. The electricity of the rubber was always contrary to that of the cylinder, but varied with different bodies.

2. Conducting Power.

It is generally stated by writers on this subject, that hot air is a conductor of electricity, because flame, or red hot substances destroy the effect of insulation. I have said at page 40, that the most intensely heated air, if unaccompanied by flame, is not a conductor. My reasons for this statement are the following.

1. I do not find that an electrified gold-leaf electrometer, or Leyden jar, is discharged when placed at but a very moderate distance in front of a large fire, unless the glass becomes so hot as to be rendered a conductor.

2. The rays of the sun, concentrated by the action of a lens, do not conduct electricity.

3. Mr. Read found that an electrified electrometer, when passed quickly into and out of an oven, so hot as to burn off one of its balls, still remained electrified. See Read on Spontaneous Electricity, p. 8, &c.

3. Preparation of Amalgam.

The amalgam described at page 52, answers exceedingly well: but I have since made it with a still less proportion of mercury with equal effect. The proportions may be two oz. of tin, four oz. of zinc, and seven oz. of mercury. The mercury must be heated to about three hundred Farenheit, before the fused metals are added to it. When the amalgam has been agitated until cool, and finely powdered, it is to be mixed with hog's-lard by trituration in a mortar; and should it at any time become hard, more lard must be added, and the trituration be repeated.

4. Ratio of Electrical Intensity, as influenced by Figure.

It has been stated at page 81, that if two spheres of unequal size are connected together and electrified, that which is smallest will evince the greatest intensity. A very satisfactory demonstration of this fact was contrived by Mr. De Luc, and requires but a simple apparatus. Two circular plates of metal, with smooth round edges, are to be provided; one of them, which I shall call A, may be ten or twelve inches diameter, it is to be supported in an horizontal position by an insulating pillar; the other which I name B, should not exceed an inch in diameter, and a twelfth of an inch thick, and must be provided with an insulating handle.

A is to be slightly electrified, either positively or negatively: B is then to be held by its insulating handle, and applied by its flat surface to any part of A; and being then removed in the same manner, and applied to the cap of an electrometer, will indicate by the divergence it produces the intensity of that part. A second contact is to be made with some other portion of A's surface, and on bringing the small plate again to the electrometer, its divergence will be either increased or diminished, and will consequently indicate the different intensity of two parts of the same conductor. In this way repeated trials may be made, and it will be found that the intensity is least at the centre of the flat surface of A; and that it gradually increases towards the circumference, being greatest at the extreme edge. A similar experiment may be made by a series of balls regularly decreasing in size and connected together by a wire; the smallest ball will exhibit the most considerable intensity.

5. ELECTRICAL BATTERY.

The practical electrician will (without great care,) find this essential part of his apparatus a source of considerable expense, from the frequent fracture of jars by spontaneous explosion. It may be therefore useful to know, that it is by no means absolutely necessary to employ jars all of the same size; for if their thickness is nearly the same, and their uncoated rims of equal extent, the effect will approximate to the proportion of the sum of the coated surfaces, though considerable difference may exist between the sizes of the jars. My friend Mr. Crosse has lately communicated to me a very extensive series of experiments on this subject; in some of which, the difference between the size of the combined jars was very considerable, yet the wire melting power (within the limits of their charging capacity,) was very nearly the proper proportion of the whole coated surface. Large green glass bottles may be coated for an electrical battery, and answer well where appearance is not considered.

6. Motion of a Pith Ball, by the discharge of a Jar.

This experiment, described at page 161, is by no means a satisfactory indication of the course of the fluid in the discharge, and from frequent repetition under circumstances of considerable variation, it appears to me that the motion is always produced by electrical attraction; for I find, that if the ball is placed on the groove as directed, at an equal distance from each of the pointed wires, it will move from the one with which the discharging rod is connected, whether that be brought in contact with a positively charged jar, or one that is negatively charged.* If the discharging rod be kept in contact with the knob of the jar, the ball remains at the wire to which it has been driven. But, if after the contact has been made, the

* Mr. Howldy has stated in Nicholson's Journal, vol. xxxiv. p. 199, that a pith ball placed in a groove at an equal distance from the ends of the two wires, does not move when the discharge is passed from one to the other; but the groove he employed "was fitted as nearly as could be to the curvature of the pith ball, and was as deep as half its diameter;" consequently the motion of the ball was prevented by friction, which the slight charge he employed could not overcome.

discharging rod is quickly withdrawn, and then brought in contact and withdrawn again, several times, the ball will move from one wire to the other, and back again repeatedly, as it would between two oppositely electrified bodies.

7. Perforation of Paper by the expansive Force of the Charge.

The truth of the reasoning applied to Mr. Symmer's experiment (at page 164,) is confirmed by a variation of it. Let six or eight sheets of tinfoil be interposed between the leaves of a quire of paper, so that the pieces of tinfoil are separated from each other, in every place, by three or four thicknesses of paper. On passing a sufficiently strong charge through the quire, its leaves will be perforated in different places, and each piece of tinfoil will have two indentations in opposite directions to each other. On the supposition that this effect is produced by contrary currents of electricity, it would be necessary to admit the existence of twice as many electric fluids as there are pieces of tinfoil; for the indentations that point in either direction are not in a line with each other. But as a further proof that it is merely the usual expansive effect of a spark at each interruption of the metallic circuit, the impression on the tinfoil is greater or less, in proportion to the number of leaves of paper that separate one piece from another; and when the sheets of tinfoil are separated by single sheets of paper only, the effect is very trivial; or if the whole thickness is inconsiderable, both the paper and the tinfoil are sometimes perforated in one direction, and that evidently from the positive to the negative.

8. Preparation of Electrical Cements.

The various cements employed in the construction of electrical apparatus are formed principally of resin, with the addition of some substances to render it more adhesive, and less brittle. Five pounds of resin, one pound of bee's-wax, one pound of red ochre, and two table spoonfuls of plaster of Paris; when melted, and well incorporated together, form a very good cement for general purposes. One that is well adapted for cementing large Voltaic batteries, and which is cheaper, may be formed of six pounds of resin, one pound of red ochre, half a pound of plaster of Paris, and a quarter of a pint of linseed oil. Other cements in great variety, more or less fusible, &c. may be formed by combining the preceding ingredients in various proportions. The ochre and plaster of Paris should be well dried, and added to the other ingredients when they are well melted.

9. On the Electrical Effects exhibited by various Bodies, after their mutual Contact.

I have lately made many experiments on this subject, which, by the aid of the new method of insulation, I was enabled to do with a degree of facility and precision before unknown.

As it was desirable to compare the electricity of the substances under examination, as obtained by their contact with the same body, two methods were employed. 1st. That of sifting them on the cap of a delicate electrometer through a fine sieve, which was thoroughly cleansed after each operation. 2d. By bringing an insulated copper-plate repeatedly in contact with extensive surfaces of them spread on a dry sheet of paper; the copper-plate being brought in contact with the condenser after every repetition of the touching, until a sufficient charge was communicated. By each process the effects increased very considerably when the substances employed were reduced to a fine powder; and it was in this way I succeeded in obtaining very distinct effects from the alkalies, by contact with a copper or a silver plate; an experiment which Sir H. Davy had before attempted with great care, but without success. The pure alkalies were broken into small pieces, and being placed in an open phial were exposed for a quarter of an hour to a moderate heat, not sufficient to fuse the alkali, which was then quickly reduced to powder in a warm and dry mortar, and immediately spread on a dry sheet of card paper, which for some time will continue to attract moisture from the alkali, as fast as the alkali receives it from the air. The whole operation was performed as rapidly as possible.

The greater effect produced in all these experiments by an increased division of the powder, renders it highly probable that they are merely varieties of the usual process of excitation.

The following substances produce negative electricity when sifted on the cap of an electrometer.

Copper, iron, zinc, tin, bismuth, antimony, nickel, black-lead, lime, magnesia, barytes, strontites, alumine, silex, brown oxide of copper, white oxide of arsenic, red oxide of lead, litharge, white lead, red oxide of iron, acetate of copper, sulphate of copper, sulphate of soda, phosphate of soda, carbonate of soda, carbonate of ammonia, carbonate of potash, carbonate of lime, muriate of ammonia, common pearl ashes, boracic acid, benzoic acid, oxalic acid, citric acid, tartaric acid, cream of tartar, oxymuriate of potash, pure potash, pure soda, resin, sulphur, sulphuret of lime, starch, orpiment, &c.

The following substances produce positive electricity when sifted on the cap of an electrometer.

Wheat-flour, oatmeal, lycopodium, quassia, powdered cardamom, charcoal of wood, sulphate of potash, nitrate of potash, acetate of lead, oxide of tin.

Hence it appears, that there are comparatively but few substances that appear positively electrified when sifted through hair, flannel, or muslin. For, in experiments made with each of these substances separately, they were found to produce similar effects.

The following table, exhibits the results of experiments of contact with a copper-plate;

the different substances are arranged in a column under the electricity they really obtain, which is contrary to that of the copper-plate.

POSITIVE.	NEGATIVE.
Lime, barytes, strontites, magnesia, pure soda, pure potash, common pearl-ashes, carbonate of potash, carbonate of soda, tartaric acid.	Benzoic acid, boracic acid, oxalic acid, citric acid, silex, alumine, carbonate of ammonia, sulphur, resin.

These experiments were several times repeated with uniform results. On the whole, they by no means favour the idea of natural electrical energy; and the result obtained with sulphur, and with resin, being similar to that produced by their friction, nearly establishes as a fact the opinion, that the contact of dissimilar bodies is in general the primary source of electrical excitation.

10. Medical Application of Voltaic Electricity.

The current of electricity produced by a Voltaic apparatus, has been applied with occasional success in some cases of palsy, rheumatism, rheumatic head-ach, deafness, and opacity of the cornea. The parts through which it is transmitted are moistened with water, and sometimes a small piece of gold or silver leaf is applied; for, as it has been before stated, the power of a moderate Voltaic apparatus is not sufficient to penetrate the dry cuticle. When applied to the eyes, a moistened piece of sponge attached to the end of a wire, is a convenient vehicle for its transmission, and a very moderate power only should be applied; for I am assured, that in some instances blindness has been produced by the injudicious application of an active battery to this delicate organ! The size of the plates is not of so much consequence as the nature of the fluid by which they are excited; I am decidedly of opinion, that a strong acid mixture should never be employed; and the contrary practice will render the action of the battery more uniform and permanent; it is highly probable that $\frac{1}{500}$th part of muriatic acid will be found the most useful proportion for most medical purposes. The practitioner may derive some useful information on the effect of different degrees of Voltaic power on the animal fluids, from Mr. Brande's paper in the Philosophical Transactions for 1809, p. 385, &c.

11. Action of the Sun's Rays on the Electric Column.

Mr. De Luc has recently observed, that the power of an electric column (as indicated by the oscillation of a pendulum,) is increased when the sun shines upon it, and his observation has been corroborated by Mr. Hausmann.* Mr. De Luc conceived, that the effect did not arise from the heat of the sun, because he had observed that a column put together with discs of paper that had been thoroughly dried, evinced very little power. It is however certain, that a moderate heat does increase the power of the column, for the bell-ringing apparatus I have, is kept in a room where there is rarely a fire, and I find that it pulsates most slowly in winter; but if a fire is made in the room, the ringing soon becomes more rapid. I took a column of a thousand series, and applied it to the cap of a gold leaf electrometer, when the temperature of my room was fifty. The gold leaves struck the sides of the glass nine times in sixty seconds. I then placed the column for ten minutes before a fire, where the thermometer rose

* Nicholson's Journal, vol, xxxvi. p. 307, &c.

to 85, and it then, on being applied to the electrometer, occasioned the gold leaves to strike the sides thirty-seven times in sixty seconds. It was afterwards removed to another part of the room until it had recovered its original temperature of 50, and it then occasioned the gold leaves of the same electrometer to strike only nine times in a minute as at first; but on placing it before the fire again for ten minutes, it produced thirty-seven strikings in a minute. Some facts nearly analogous to this, which demonstrate a very remarkable influence of temperature on the electricity produced by the contact of different bodies, and on the action of the Voltaic apparatus; are detailed in a Memoir read before the French Institute, Sept. 23d. 1811. By J. P. Dessaignes. See the Journal de Physique for 1811, vol. lxxiii, p. 230. And at page 417 of the same volume, a supplement to the Memoir, in a letter to the editor, from M. Dessaignes.

12. Production of the Electrical Oxides.

The oxidation of metals by electricity, described at page 182, is usually effected by rather high charges of a moderate sized battery, and under such circumstances some of the jars are very frequently broken. I have found, that by increasing the extent of the battery, more moderate charges are sufficient; and my friend, Mr. Crosse, has observed the same circumstance in the employment of his very extensive and powerful apparatus.

The law first noticed by Mr. Brooke with regard to the fusion of metals, seems also to obtain when they are oxidated, for I find that a battery of 40 square feet of coated surface, charged to 10 grains, will oxidate the same quantity of gold wire as a battery of 20 feet, charged to 20 grains, and the chance of fracture is much less with the lowest charge.

It was by the employment of a very extensive battery in this way, that the specimens of the oxides of gold and copper, with which a few copies of this work are illustrated, were produced.

The figures impressed on glass and paper by the electrical oxides, vary materially, even when produced under apparently similar circumstances, and in a very considerable number of experiments, I have never obtained two specimens exactly alike.

T. BENSLEY, PRINTER,
Bolt Court, Fleet Street.

PLATE I.

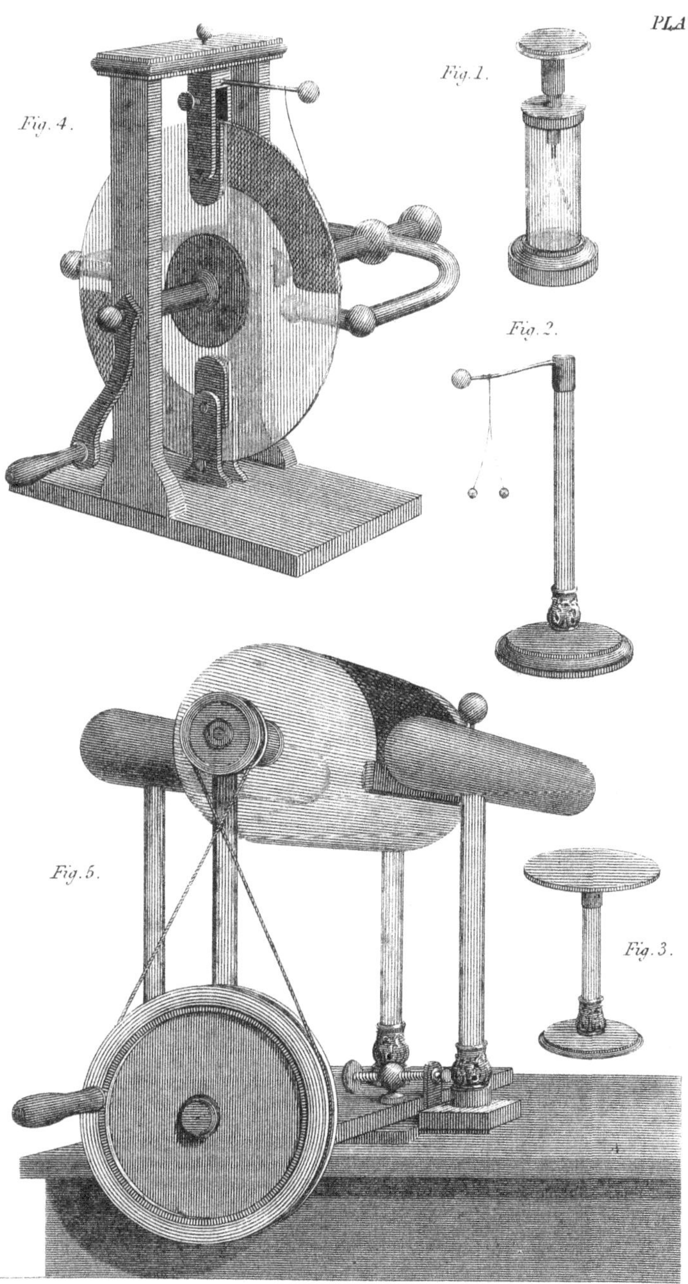

Published by Longman & Co, Feb.y 1824. Lowry sculp.

PLATE II.

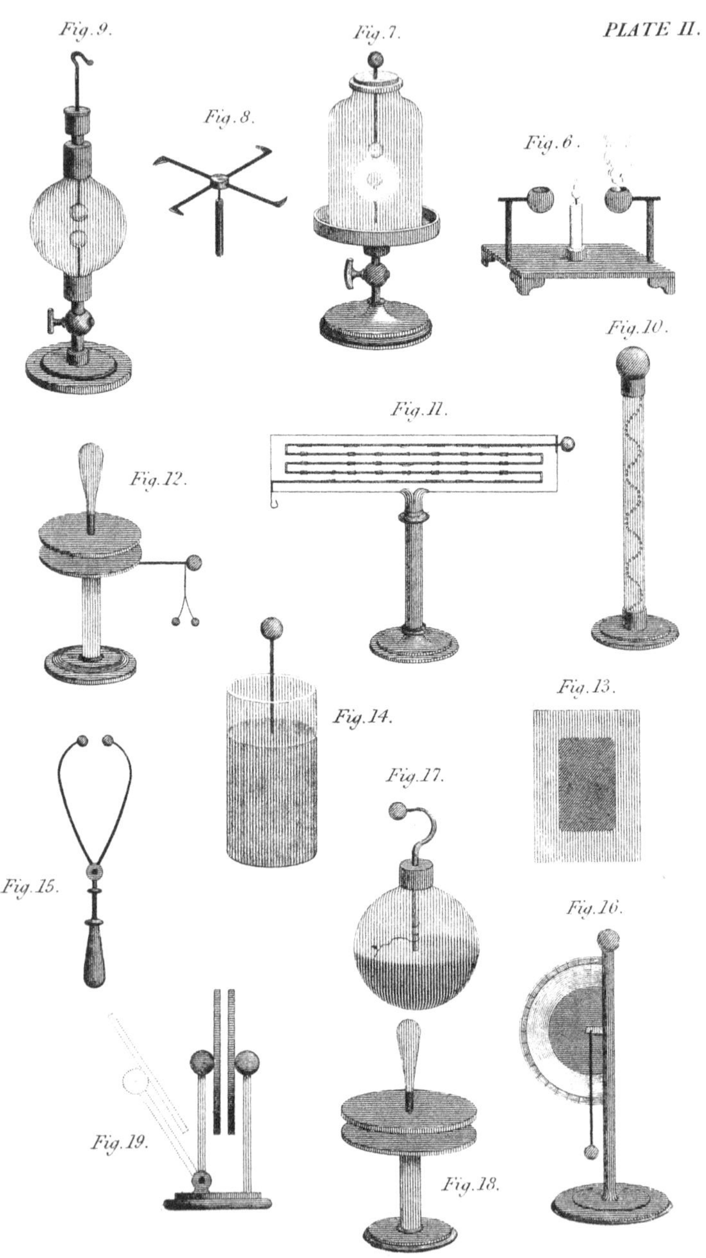

Published by Longman & Co. Feb.y 1814.

Lowry sculp.

PLATE III.

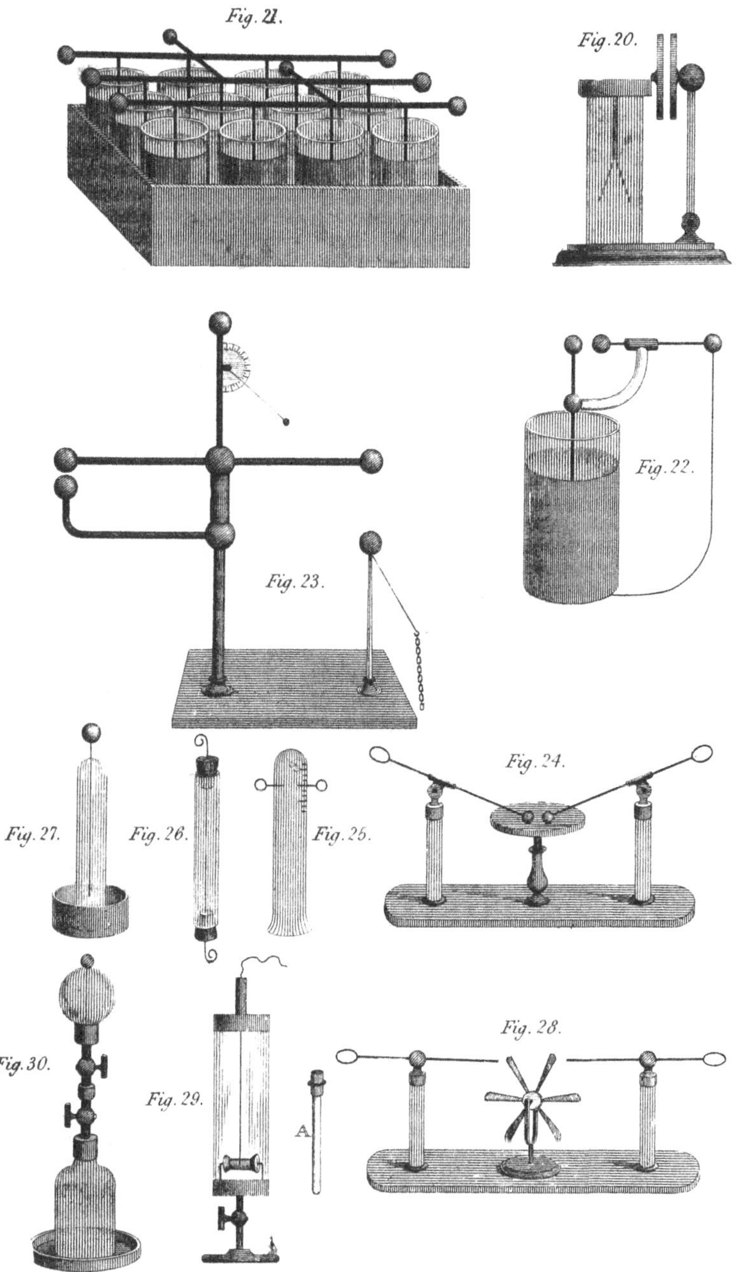

Published by Longman & Co. Feb.ʸ 1814.

Lowry sculp.

PLATE IV.

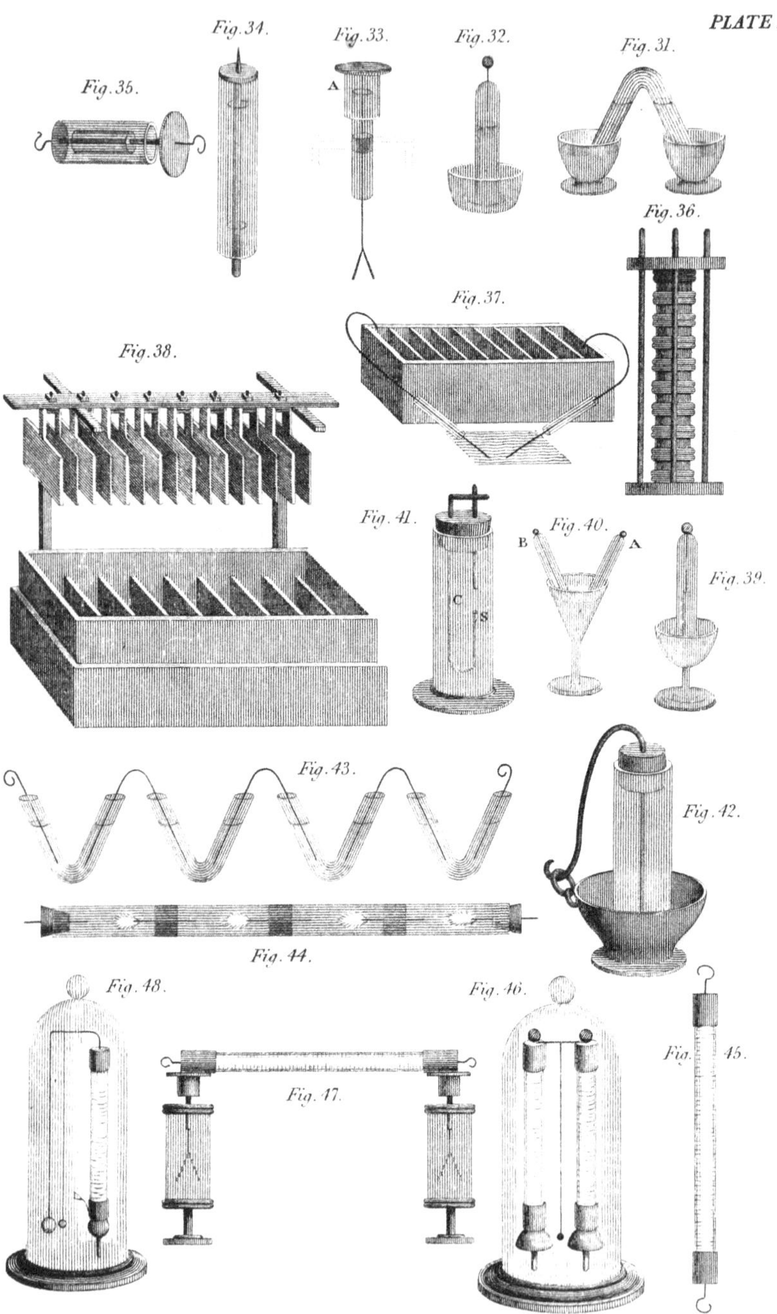

Published by Longman & Co. Feby 1814. Lowry sculp.

Printed in the United States
By Bookmasters